Measuring *in vivo* Oxidative Damage

A Practical Approach

Edited by

J. Lunec
Molecular Toxicity Group, University of Leicester, UK

and

H. R. Griffiths
Pharmacentical Sciences, Aston University, UK

JOHN WILEY & SONS, LTD

Chichester · New York · Weinheim · Brisbane · Toronto · Singapore

2000

Other Wiley Editorial Offices

John Wiley & Sons, Inc., 605 Third Avenue,
New York, NY 10158-0012, USA

WILEY-VCH Verlag GmbH, Pappelallee 3,
D-69469 Weinheim, Germany

Jacaranda Wiley Ltd, 33 Park Road, Milton,
Queensland 4064, Australia

John Wiley & Sons (Asia) Pte Ltd, Clementi Loop #02-01,
Jin Xing Distripark, Singapore 129809

John Wiley & Sons (Canada) Ltd, 22 Worcester Road,
Rexdale, Ontario M9W IL1, Canada

Library of Congress Cataloging-in-Publication Data

Measuring in vivo oxidative damage: a practical approach / edited by J. Lunec.
   p. cm.
Includes bibliographical references and index.
ISBN 0-471-81848-8 (alk. paper)
   1. Active oxygen–Physiological effect–Research–Methodology. 2. Free radicals
(Chemistry)–Pathophysiology–Research–Methodology. 3. Oxidation,
Physiological–Research–Methodology. 4. DNA–Oxidation–Research–Methodology. 5.
Proteins–Oxidation–Research–Methodology. I. Lunec, J.

RB170 .M43 2000
616.07–dc21

99-047679

British Library Cataloguing in Publication Data

A catalogue record for this book is available from the British Library

ISBN 0 471 81848 8

Typeset in 10/13pt Times by Kolam Information Services Pvt Ltd, Pondicherry, India
Printed and bound in Great Britain by Antony Rowe Ltd, Chippenham, Wiltshire
This book is printed on acid-free paper responsibly manufactured from sustainable
forestry, in which at least two trees are planted for each one used for paper production

Measuring
in vivo Oxidative Damage
A Practical Approach

Contents

Part IV   Molecular-based Assays   **105**

9   A ^{32}P-Postlabelling Protocol to Measure Oxidative DNA Damage   107
George D.D. Jones and Michael Weinfeld

10   Mapping Reactive Oxygen-Induced DNA Damage at Nucleotide Resolution   125
Henry Rodriguez and Steven A. Akman

List of Contributors

Steven A. Akman Department of Cancer Biology, Comprehensive Cancer Center of Wake Forest University, Winston-Salem, North Carolina 227157, USA

Ruth J. Bevan Division of Chemical Pathology, Glenfield Hospital NHS Trust, University of Leicester, Groby Road, Leicester, LE3 9QP, UK

Jean Cadet Département de Recherche Fondamentale sur la Matière Condensée, SCIB/Laboratoire des Lésions des Acides Nucléiques, CEA/Grenoble, F-38054 Grenoble Cedex 9, France

Andrew R. Collins Rowett Research Institute, Greenburn Road, Bucksburn, Aberdeen, AB21 9SB, UK

Marcus Cooke Division of Chemical Pathology, Centre for Mechanisms of Human Toxicity, University of Leicester, Leicester, LE1 8HN, UK

Thierry Douki Département de Recherche Fondamentale sur la Matière Condensée, SCIB/Laboratoire des Lésions des Acides Nucléiques, CEA/Grenoble, F-38054 Grenoble Cedex 9, France

Bernd Epe Institute of Pharmacy, University of Mainz, Staudinger Weg 5, D-55099 Mainz, Germany

Mark D. Evans Division of Chemical Pathology, Centre for Mechanisms of Human Toxicity, University of Leicester, Leicester, LE1 8HN, UK

Helen R. Griffiths Pharmaceutical Sciences, Aston University, Aston Triangle, Birmingham, B4 7ET, UK

Karl Herbert Division of Chemical Pathology, Centre for Mechanisms of Human Toxicity, University of Leicester, Leicester, LE1 9HN, UK

George D.D. Jones   Biomolecular Damage Group, Centre for Mechanisms of Human Toxicity, University of Leicester, Leicester, LE1 9HN, UK

Steffen Loft   Institute of Public Health, Panum Institute, University of Copenhagen, Blegdamsvej 3, DK-2200 Copenhagen, Denmark

Joe Lunec   Molecular Toxicity Group, University of Leicester, Lancaster Road, Leicester, LE1 9HN, UK

Simon R.J. Maxwell   Clinical Pharmacology Unit, The University of Edinburgh, Western General Hospital, Edinburgh, EH4 2XU, UK

Michael Pflaum   Institute of Pharmacy, University of Mainz, Staudinger Weg 5, D-55099 Mainz, Germany

Henrik E. Poulsen   Department of Clinical Pharmacology Q7642, Rigshospitalet University Hospital Copenhagen, 20 Tagensvej, DK-2200 Copenhagen, Denmark

Jean-Luc Ravanat   Département de Recherche Fondamentale sur la Matière Condensée, SCIB/Laboratoire des Lésions des Acides Nucléiques, CEA/Grenoble, F-38054 Grenoble Cedex 9, France

Henry Rodriguez   Biotechnology Division, National Institute of Standards and Technology, Gaithersburg, Maryland 20899, USA

Allan Weimann   Department of Clinical Pharmacology Q7642, Rigshospitalet University Hospital Copenhagen, 20 Tagensvej, DK-2200 Copenhagen, Denmark

Michael Weinfeld   Department of Radiobiology, Cross Cancer Institute, 11560 University Avenue, Edmonton, Alberta, T6G 1Z2, Canada

Preface

Fifteen years ago the field of free radicals was just beginning to launch into a new era of importance in pathology. Free radicals have been implicated in a number of major disease processes, including chronic inflammation, carcinogenesis and atherosclerosis. Many of the recent developments, such as the detection and measurement of oxygen free radicals *in vivo*, have borne the brunt of much criticism because clinicians have been led to believe that (i) it is not possible to measure oxidative damage *in vivo* as free radicals are too reactive; (ii) even though you can measure the reaction products of free radicals (i.e. protein, lipid and DNA damage) there is too much artefactual oxidation generated to be able to measure, accurately, the consequences of free radical damage, *in vivo*.

When clinicians require to measure an analyte, it is usually detected in either serum, plasma, urine or peripheral blood cells. Only very rarely do they use biopsy material. This need for simple, accurate and non-invasive techniques that have rapid throughput is the hallmark of the clinical biochemistry/pathology laboratory. The issue of quality assurance is also an important area to consider. The cynic might say that, according to the latter criteria, the measurement of free radical activity is not yet applicable to clinical material.

With these issues in mind we have arranged this short handbook to bring together leading experts in the field of oxidative damage and scientists who have more experience of the practical problems associated with measuring oxidative damage *in vivo*. This is by no means a comprehensive treatment of all the methods that are available, but is designed to help the clinical scientist in evaluating which of the most popular methods is the best to use for their particular clinical problem. The methods represented are the ones most commonly used and deemed most appropriate to apply to clinical material. Many of the authors are involved in national and international quality assurance programmes in order to establish these assays routinely in clinical research laboratories. Others are young scientists from clinical biochemistry laboratories who make a special study of the practical problems, associated with methods they use. These problems, and some solutions, are highlighted throughout the text.

J. Lunec
H.R. Griffiths

Part I Chromatographic Procedures

1 Lipid Peroxidation Determination by HPLC

Ruth. J. Bevan

INTRODUCTION

In recent years, evidence has been accumulating that lipid peroxidation and its associated processes are linked to a variety of pathological conditions, which include atherosclerosis, carcinogenesis, diabetes, inflammation and ischaemia reperfusion injury (Yagi, 1982; Steinberg, 1987; Halliwell and Gutteridge, 1989). Hydroperoxide formation from unsaturated phospholipids, cholesterol, triacylglycerols and other lipids can be induced by activated oxygen species, and can be summarized by the following reactions:

$$LH + R^{\bullet} \longrightarrow L^{\bullet} \tag{1.1}$$

$$L^{\bullet} + O_2 \longrightarrow LOO^{\bullet} \tag{1.2}$$

$$LOO^{\bullet} + LH \longrightarrow L^{\bullet} + LOOH \tag{1.3}$$

Lipids (LH) are susceptible to attack by radical species ($R^{\bullet}$; especially hydroxyl radical, $OH^{\bullet}$) which causes abstraction of H atoms (Eq. 1.1) and forms a lipid radical ($L^{\bullet}$). Lipid peroxide radicals ($LOO^{\bullet}$) are subsequently formed following the addition of oxygen (Eq. 1.2). $LOO^{\bullet}$ can then react with additional lipid (Eq. 1.3), again forming the lipid radical ($L^{\bullet}$) and lipid hydroperoxide (LOOH). Since the peroxide products of the above reactions can themselves act as initiators of lipid peroxidation, the reaction is autocatalytic.

Lipid hydroperoxides are the major initial product; they decompose to form secondary products of aldehydes such as malondialdehyde (MDA), 4-hydroxynonenal (4-HNE) and volatile hydrocarbons. Although it is highly desirable to detect and identify lipid hydroperoxides and associated products in biological tissues, it has proved difficult, due to their trace concentrations, instability and diversity (Akasaka *et al.*, 1993b).

Measuring in vivo *Oxidative Damage: A Practical Approach*. Edited by J. Lunec and H. R. Griffiths. © 2000 by John Wiley & Sons, Ltd. ISBN 0 471 81848 8.

Many of the methods developed to date measure either the primary hydroperoxides or secondary aldehydic products of lipid oxidation. However, these methods are complicated when applied to the *in vivo* situation. In addition, some of the more simple widely used spectrophotometric assays such as the traditional thiobarbituric acid (TBA) test (Draper and Hadley, 1990) or measurement of conjugated dienes (Recknagel and Glende, 1984) lack absolute specificity and sensitivity. Therefore, in recent years much emphasis has been placed on establishing sensitive, specific and robust assays for measurement of lipid hydroperoxides and associated products.

HPLC DETERMINATION OF LIPID HYDROPEROXIDES

HPLC (high performance liquid chromatography) is a popular tool in the measurement of lipid hydroperoxides. It has the advantage that different classes of lipid hydroperoxides can be extracted, and separated from a variety of sources such as tissues, plasma and lipoproteins. Several different systems have been employed for the high sensitivity detection and quantitation of LOOHs, following separation by HPLC. These include: measurement of chemiluminescence generated in a microperoxidase/isoluminol postcolumn reaction system (Frei *et al.*, 1988); detection using an iron/thiocyanate assay with spectrophotometric detection (Mullertz *et al.*, 1990); fluorometric detection following postcolumn reaction with diphenyl-1-pyrenylphosphine (Akasaka *et al.*, 1993a); and electrochemical detection (Korytowski *et al.*, 1993).

In this chapter, two of the methods we employ in our laboratories for the measurement of *in vivo* lipid peroxidation are described. The first is an HPLC method employing chemiluminescence detection for lipid hydroperoxides. The second method is an improved TBA assay that utilizes HPLC separation to overcome problems of specificity commonly associated with the traditional spectrophotometric assay.

Protocol 1: Determination of lipid hydroperoxides by HPLC with chemiluminescence detection

The first method is routinely used in our laboratories to measure the concentration of phospholipid hydroperoxides (particularly phosphatidylcholine hydroperoxides, PC-OOH) associated with plasma low density lipoprotein (LDL). Oxidation of LDL is of relevance to the development of coronary heart disease (Goldstein *et al.*, 1979). The method can, however, be used to measure PC-OOH of both plasma and tissue samples. The chromatographic conditions given in Protocol 1 have been optimized specifically to determine PC-OOH; however, conditions can be altered to determine other classes of lipid hydroperoxides, e.g. triacylglycerol and cholesterol ester hydroperoxides (Akasaka *et al.*, 1993b) and cholesterol hydroperoxides (Korytowski *et al.*, 1993).

The detection system used by us is based on the specific chemiluminescence reaction of luminol/microperoxidase (heme fragment of cytochrome *c*) with hydroperoxides. It is specific for the hydroperoxide and has reported picomole sensitivity (Yamamoto *et al.*, 1990). We have found that a luminol/microperoxidase combination gives greatest sensitivity in our system, but other authors have used an isoluminol/microperoxidase combination. The mechanism is thought to be as follows:

$$LOOH + microperoxidase \longrightarrow LO^{\bullet} \qquad (1.4)$$

$$LO^{\bullet} + luminol\ (QH^-) \longrightarrow LOH + semiquinone\ radical\ (Q^{\bullet}) \qquad (1.5)$$

$$Q^{\bullet} + O_2 \longrightarrow quinone\ (Q) + O_2^{\bullet-} \qquad (1.6)$$

$$Q^{\bullet} + O_2^{\bullet-} \longrightarrow luminol\ endoperoxide \longrightarrow light \qquad (1.7)$$

$$(\lambda_{max} 430\ nm)$$

A schematic of the HPLC system used is shown in Figure 1.1.

Equipment and Reagents
- Cooled bench-top centrifuge
- Bench-top ultracentrifuge
- Nitrogen heating block
- HPLC system with Spherisorb Phenyl column and chemiluminescence detector
- Sonicating water bath
- EDTA blood collection tubes
- Stoppered glass tubes
- Butylated hydroxytoluene (BHT)
- HPLC grade chloroform and methanol
- L-Phosphatidylcholine (Sigma, P3556)
- 50 mM Sodium borate buffer, pH 10.0

- Luminol (5-amino-2,3-dihydro-1,4-phthalazinedione, Sigma A8511)
- Microperoxidase (MP-11, Sigma M6756)

LDL isolation and extraction

1. Collect blood sample into EDTA (1 mg/ml) and centrifuge $1000 \times g$ for 20 min at 4°C to obtain plasma. Samples should be kept on ice for no longer than 1 h before use.
2. Isolate LDL fraction using method of single vertical spin density gradient ultracentrifugation (Chung *et al.*, 1986). Wash fraction by re-centrifugation under the same conditions in order to remove traces of albumin.
3. Extract lipids from LDL sample using 2 volumes of 2:1 (v/v) chloroform:methanol containing 0.01% BHT to prevent autoxidation. Vortex samples for 30 s and collect chloroform layer by centrifugation at $1000 \times g$ for 10 min at 4°C.
4. Dry chloroform layer under nitrogen at 45°C in a heating block. Dried samples may be stored under nitrogen at −70°C for up to 1 week.

Preparation of phosphatidylcholine hydroperoxide standard

1. Dissolve phosphatidylcholine standard in methanol to give 20 mg/ml solution, and place in a stoppered glass tube.
2. Incubate at 37°C for 72 h (longer than this may cause hydroperoxides to decompose).
3. Assess concentration of lipid hydroperoxides by spectrophotometric analysis of potassium iodide reduction (El-Saadani *et al.*, 1989). Store standard at −20°C for up to 2 months and re-measure hydroperoxide concentration at regular intervals.

HPLC analysis

1. Equilibrate Spherisorb Phenyl column (250×4.6 mm, Phenomenex) with 100% methanol, for minimum of 30 min.
2. Prepare chemiluminescence reagent comprising 70:30 (v/v) methanol:50 mM sodium borate buffer, pH 10.0, containing 1 mM luminol and 10 μg/ml microperoxidase. Place in a dark bottle and degas for 30 min in a sonicating water bath. Fol-

lowing sonication, stand reagent for approximately 30 min prior to use.[a]

3. Pump eluent and chemiluminescence reagent at 1 ml/min from pumps 1 and 2, respectively (see Figure 1.1). Monitor baseline from chemiluminescence detector (Soma model S-3400) for approximately 10 min to assess background noise.

4. Reconstitute dried LDL fraction using 200 μl 1:1 (v/v) chloroform:methanol containing 0.01% BHT to prevent autoxidation.

5. Inject all the sample immediately onto column and monitor peak at approximately 7 min retention (Figure 1.2).

6. Quantitate peak obtained by standard curve prepared by injecting standard phosphatidylcholine hydroperoxide in range 0–250 μM.[b]

[a] Reagent is stood for 30 min to allow inherent chemiluminescence to subside.
[b] Limit of detection found to be 4 pmol using a 200 μl injection volume; standard curve found to be linear to concentration of 1 μM PC-OOH.

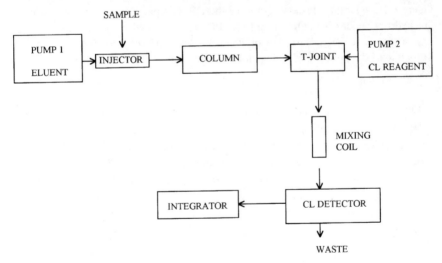

Figure 1.1 HPLC system for quantitation of lipid hydroperoxides using chemiluminescence detection. All parameters are as described in Protocol 1. T-joint, three-way mixing tee; CL reagent, chemiluminescence reagent; CL detector, chemiluminescence detector. (*Note*: Unoxidized lipid species can be monitored simultaneously by inserting a UV detector (210 nm) after the column.)

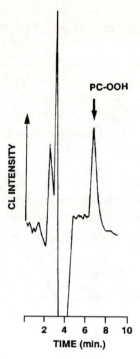

Figure 1.2 A sample chromatograph of the HPLC separation of phosphatidylcholine hydroperoxide (PC-OOH) from LDL. PC-OOH is eluted at 7.0 min, and the peak shown represents a concentration of 70 nm PC-OOH. (Note quenching of signal at 4 min due to the presence of endogenous antioxidants in the sample.)

One of the drawbacks of the chemiluminescence detection system for use with biological samples is that the signal is quenched by antioxidants. This problem can be overcome by optimization of HPLC conditions to separate antioxidants from lipid hydroperoxide peaks. The negative peak caused by elution of antioxidants contained within LDL extracts can be seen in Figure 1.2. When setting up the assay it is also important to consider the background noise caused by inherent chemiluminescence within the reagent, which can decrease the sensitivity of the assay. The signal to noise ratio can be maximized by optimizing the following parameters: (i) concentration of microperoxidase; (ii) concentration of luminol/isoluminol; (iii) mixing coil length; (iv) retention time of lipid hydroperoxides.

In our laboratories the assay described above has proved to be specific, sensitive and robust, with sample intra-and inter-assay variations of 7.0 and

6.0% CV, respectively. The recovery of phosphatidylcholine hydroperoxide from LDL fractions using the method in Protocol 1 was typically 80% (quadruplicate analysis of 125 nM PC-OOH spike in normal plasma). Using this assay we have measured LDL lipid hydroperoxide concentrations of 50 ± 20 nM PC-OOH in healthy subjects ($n = 20$) which compares favourably with previously reported data (Miyazawa et al., 1990). In addition, we have measured PC-OOH concentrations in LDL isolated from hyperlipidemic subjects of 68 ± 18 nM ($n = 23$), which is significantly higher ($p < 0.01$) when compared with healthy subjects. This may indicate increased oxidative stress in individuals with raised lipid concentrations, leading to increased susceptibility to development of atherosclerosis.

Protocol 2: Measurement of lipid hydroperoxides using an HPLC-based thiobarbituric acid assay

One of the most popular assays of lipid peroxidation is by measurement of the aldehyde MDA, which is formed by the breakdown of lipid hydroperoxide. This is carried out using the classical TBA test. In the assay, samples are reacted with TBA by heating under acidic conditions. Pre-formed MDA in the sample reacts with TBA to form TBA–MDA adduct consisting of two molecules of TBA and one of MDA (Figure 1.3). This adduct has a very high absorption at 532 nm and a strong fluorescence at excitation 536 nm and emission 555 nm, which makes the assay very sensitive. However, the TBA assay does suffer from interference from a number of substances unrelated to lipid peroxidation, which react with TBA to form chromogens with absorbency around 530–535 nm. Although detection by fluorescence can improve specificity, these interferences have led to the validity of the TBA assay being questioned, especially when applied to complex biological systems such as plasma (Gutteridge, 1986).

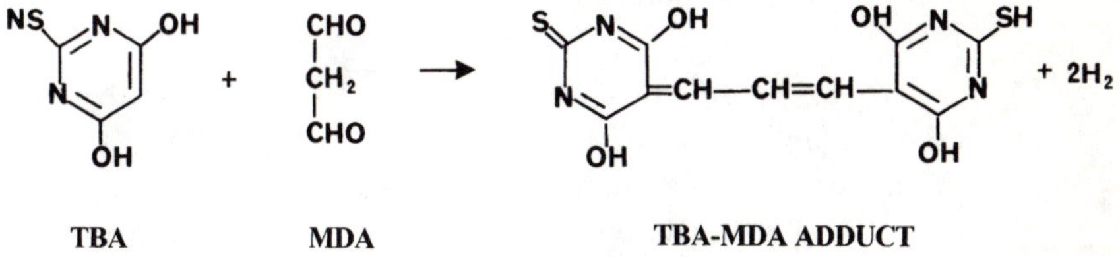

Figure 1.3 Schematic of the reaction of thiobarbituric acid (TBA) with malondialdehyde (MDA). The TBA–MDA adduct is a coloured product detected by either absorbance at 532 nm, or fluorescence with Ex. 536 nm and Em. 555 nm.

The TBA assay can be made more specific by using HPLC to separate and directly measure the TBA–MDA adduct, and a number of assays have been described (Bird *et al.*, 1983; Therasse and Lemonnier, 1987; Wong *et al.*, 1987). All these methods give lower values for tissue and plasma thiobarbituric acid substances (TBARs) than the traditional spectrophotometric assay but can be fairly intensive, requiring complex sample preparation or organic sample extractions in order to improve sensitivity and peak separations. In our laboratories we use the method developed by Young and Trimble (1991), which has the advantage of requiring only isocratic HPLC separation with

Equipment and reagents
- EDTA blood collection tubes
- Cooled bench-top centrifuge
- Screw-capped eppendorf tubes
- Heated water bath
- HPLC system with Techopak C18 column and fluorescence detector
- TEP (1,1,3,3-tetraethoxypropane; Sigma, T9889)
- Butylated hydroxytoluene (BHT)
- Phosphoric acid
- Thiobarbituric acid (BDH-30408 4Y)
- HPLC grade absolute ethanol and methanol

Plasma preparation
1. Collect blood into EDTA (1 mg/ml) and centrifuge immediately at $1000 \times g$ for 20 min at 4°C.
2. Separate plasma and store in aliquots at −70°C until analysis.
3. Samples should be processed as soon as possible, but are stable for up to 4 months at −70°C.

Reaction with thiobarbituric acid
1. Prepare duplicate standards in the range 0–1.25 μM MDA[a] by dilution of a stock solution of TEP[b] with absolute ethanol.
2. Place 100 μl of standard or sample in a screw-capped eppendorf tube and add 10 μl of 0.2% BHT (in ethanol) and 600 μl of 0.46 M phosphoric acid. Vortex mix and leave to stand at room temperature for 10 min.

3. Add 200 μl of 0.6% TBA to all tubes, vortex mix, and heat for 30 min at 90°C in a water bath.
4. Cool tubes on ice. Take 400 μl of sample or standard and add 720 μl of methanol and 80 μl of 1 M NaOH in order to neutralize the acid and precipitate protein.
5. Spin down samples and standards in a microfuge for 5 min and remove supernatant.

HPLC analysis
1. Equilibrate Techopak C18 (30 cm × 3.9 mm; HPLC Technology) column with eluent consisting 35:65 (v/v) methanol:25 mM phosphate buffer, pH 6.4, for 60 min at 1 ml/min.
2. Set fluorescence detector at Excitation 536 nm and Emission 555 nm.
3. Inject 40μl of sample or standard, and measure peak at 6 min retention (Figure 1.4).
4. Quantitate sample using peak areas obtained from the standards.

[a] Standard curve has been shown to be linear up to 5 μM MDA. Limit of detection calculated as 4 pmol MDA on the column.
[b] TEP yields equimolar concentrations of MDA under the conditions of the reaction. Actual concentration of MDA obtained can be determined using the molar extinction coefficient of the TBA–MDA adduct at 532 nm (1.56 × 10^5 1 cm^{-1}mol^{-}1).

fluorometric detection, and does not require complex sample preparation or extraction.

Measurement of MDA concentration in plasma and other biological samples using the TBA assay continues to be the most popular marker of lipid peroxidation. However, the normal range reported for TBARs has differed widely, depending on the type of assay used. Reported values using HPLC-modified TBA assays are lower than the conventional spectrophotometric versions. We have obtained values in healthy subjects of 0.23 ± 0.08 μM MDA, which compare well with other published values (Knight *et al.*, 1987; Wong *et al.*, 1987; Young and Trimble, 1991). The method has proved to be robust, with sample intra- and inter-assay variations of 7.6 and 9.1% CV, respectively.

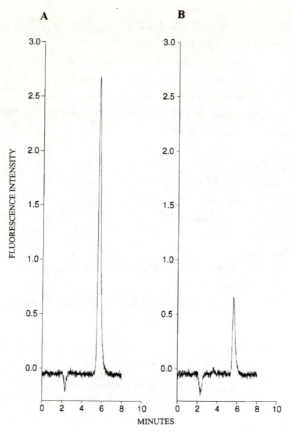

Figure 1.4 Sample chromatographs of the HPLC separation of TBA–MDA adduct. (A) Standard representing $1.25\,\mu$M MDA; (B) plasma sample containing $0.31\,\mu$M MDA. Detection was by fluorescence with Ex. 536 nm and Em. 555 nm.

CONCLUSIONS

Lipid peroxides have been reported to have a role in a number of disease states. The need for specific and sensitive assays with which to quantitate *in vivo* lipid peroxidation is therefore becoming increasingly important. HPLC has a major role to play in the analysis of LOOHs, offering several advantages over other techniques. HPLC requires a minimum of sample work-up, thereby decreasing potential artefactual oxidation of lipids, and can be utilized in analysis of a variety of biological samples. Separation of LOOHs into specific lipid classes can also be achieved, which is important for establishing the potential role of individual lipid classes in the development of disease. Finally, HPLC can be used with existing and widely used assays of LOOHs to increase specificity and sensitivity, thereby possibly bringing greater uniformity to reported LOOH values.

REFERENCES

Akasaka, K., Ohrui, H. and Meguro, H. (1993a) Determination of triacylglycerol and cholesterol ester hydroperoxides in human plasma by high-performance liquid chromatography with fluorometric postcolumn detection. *J. Chrom. Biomed. Applications*, **617**, 205–211.

Akasaka, K., Ohrui, H. and Meguro, H. (1993b) Normal-phase high-performance liquid chromatography with a fluorimetric postcolumn detection system for lipid hydroperoxides. *J. Chrom.*, **628**, 31–35.

Bird, R.P., Hung, S.S.O., Hadley, M. *et al.* (1983) Determination of malonaldehyde in biological materials by high-pressure liquid chromatography. *Anal. Biochem.*, **128**, 240–244.

Chung, B.H., Segrest, J.P., Ray, M.J. *et al.* (1986) Single vertical spin density gradient ultracentrifugation. *Meth. Enzymol.*, **128**, 181–209.

Draper, H. and Hadley, M. (1990) Malondialdehyde determination as index of lipid peroxidation. *Meth. Enzymol.*, **186**, 421–431.

El-Saadani, M., Esterbauer, H., El-Sayed, M. *et al.* (1989) A spectrophotometric assay for lipid peroxides in serum lipoproteins using a commercially available reagent. *J. Lipid Res.*, **30**, 627–630.

Frei, B., Yamamoto, Y., Niclas, D. *et al.* (1988) Evaluation of an isoluminol chemiluminescence assay for detection of hydroperoxides in human blood plasma. *Anal. Biochem.*, **175**, 120–130.

Goldstein, J.L., Ho, Y.K., Basu, S.K. *et al.* (1979) Binding site on macrophages that mediates uptake and degradation of acetylated low density lipoprotein producing massive cholesterol deposition. *Proc. Natl. Acad. Sci. USA*, **76(1)**, 333–337.

Gutteridge, J.M.C. (1986) Aspects to consider when detecting and measuring lipid peroxidation. *Free Rad. Res. Comms.*, **1**, 173–184.

Halliwell, B. and Gutteridge, J.M.C. (eds) (1989) In *Free Radicals in Biology and Medicine*, Chp. 4 pp. 188–276. Oxford/Clarendon Press, New York.

Knight, J.A., Smith, S.E., Kinder, V.E. *et al.* (1987) Reference intervals for plasma lipoperoxides: age-, sex- and specimen-related variations. *Clin. Chem.*, **33**, 2289–2291.

Korytowski, W., Bachowski, G.J. and Girotti, A.W. (1993) Analysis of cholesterol and phospholipid hydroperoxides by high-performance liquid chromatography with mercury drop electrochemical detection. *Anal. Biochem.*, **213**, 111–119.

Miyazawa, T., Fujimoto, K. and Oikawa, S. (1990) Determination of lipid hydroperoxides in low density lipoprotein from human plasma using high performance liquid chromatography with chemiluminescence detection. *Biomed. Chrom.*, **4**, 131–134.

Mullertz, A., Schmedes, A. and Holmer, G. (1990) Separation and detection of phospholipid hydroperoxides in the low nanomolar range by a high performance liquid chromatography/ironthiocyanate assay. *Lipids*, **25**, 415–418.

Recknagel, R. and Glende, E. Jr (1984) Spectrophotometric detection of lipid conjugated dienes. *Meth. Enzymol.*, **105**, 331–337.

Steinberg, D. (1987) Lipoproteins and the pathogenesis of atherosclerosis. *Circulation*, **76**, 508–514.

Therasse, J. and Lemonnier, F. (1987) Determination of plasma lipoperoxides by high-performance liquid chromatography. *J. Chrom.*, **413**, 237–241.

Wong, S., Knight, J., Hopfer, S. *et al.* (1987) Lipoperoxides in plasma as measured by liquid-chromatographic separation of malondialdehyde–thiobarbituric acid adduct. *Clin. Chem.*, **33**, 214–220.

Yagi, K. (ed). (1982) *Lipid Peroxides in Biology and Medicine*. Academic Press, New York.

Yamamoto, Y., Frei, B. and Ames, B.N. (1990) Assay of lipid hydroperoxides using high-performance liquid chromatography with isoluminal chemiluminescence detection. *Meth. Enzymol.*, **186**, 371–380.

Young, I.S. and Trimble, E.R. (1991) Measurement of malondialdehyde in plasma by high performance liquid chromatography with fluorimetric detection. *Ann. Clin. Biochem.*, **28**, 504–508.

2 Measurement of Oxidized Bases in DNA and Biological Fluids by Gas Chromatography Coupled to Mass Spectrometry

Thierry Douki, Jean-Luc Ravanat and Jean Cadet

INTRODUCTION

A method based on the use of gas chromatography coupled to mass spectrometry has been optimized to monitor the formation of oxidized bases and nucleosides within cellular DNA and biological fluids. A critical HPLC prepurification allows the separation of the targeted compound from its base precursor. Immunoaffinity prepurification has also been developed for 7, 8-dihydro-8-oxoguanine (8-oxoGua) (Ravanat *et al.*, 1995; Douki *et al.*, 1996a). This is aimed at preventing the artefactual oxidation of the overwhelming normal DNA components during the derivatization step, which has been shown to lead to overestimated values (Ravanat *et al.*, 1995; Douki *et al.*, 1996a). The assay can also be applied to biological fluids such as urine. A prepurification has also to be performed due to the wide variety of compounds present in solution (Bianchini *et al.*, 1996). In addition, the conditions of acidic hydrolysis of DNA have been optimized to allow a quantitative release of the modified bases (Douki *et al.*, 1996b; Douki *et al.*, 1997).

The following points should be noted:

- Major efforts have been made to minimize the artefactual oxidation of DNA bases during the extraction and subsequent work-up. This is based on an HPLC-electrochemical detection of 7,8-dihydro-8-oxo-2'-deoxy-

---

Measuring in vivo *Oxidative Damage: A Practical Approach*. Edited by J. Lunec and H. R. Griffiths. © 2000 by John Wiley & Sons, Ltd. ISBN 0 471 81848 8.

guanosine (8-oxodGuo). The basic strategy involves the addition of metal chelators, including EDTA and desferroxamine

- Isotopically labelled [M+3] or [M+4] derivatives of the compounds of interest are used as internal standards. They are added to the samples as early as possible in the protocol, i.e. prior to the hydrolysis. This allows correction for all possible loss of material, especially during the prepurification step.

- The HPLC prepurification steps prior to the silylation have to be carried out on dedicated columns on which only low amounts (< 1 nmol) of the targeted molecules are injected; otherwise a constant release of tiny amounts of the product occurs, leading to inaccurate data when analyzing trace amounts. Practically, the fractions to be collected can be determined by injecting 500 pmol of the compounds of interest and collecting fractions in a blind manner every minute. The resulting solutions are analyzed by GCMS.

- It should be emphasized that this protocol has been successfully applied to isolated DNA. When applied to large amounts of liver DNA (300 μg), the sensitivity of the MS detector allowed only the detection of 8-oxoGua. As far as oxidized pyrimidines are concerned, the maximum value for their level in liver DNA is in the range of 5 lesions per 10^6 nucleobases.

REAGENTS

- 2'-Deoxyguanosine (dGuo), 5-(hydroxymethyl)uracil (5-HMUra), 5-formyluracil (5-ForUra), Tris, desferroxamine mesylate, Triton-100, RNase A, RNase T1 and CNBr-activated sepharose 4B are obtained from Sigma (St Louis, Missouri).

- Sodium chloride and EDTA are purchased from Merck (Darmstadt, Germany).

- β-Mercaptoethanol, 5-hydroxyuracil (5-OHUra, isobarbituric acid), N-tertbutyldimethylsilyl-N-methyl-trifluoroacetamide and N-bis(trimethylsilyl) trifluoroacetamide are from Fluka (Buchs, Switzerland).

- 8-OxoGua is obtained from Chemical Dynamics Corporation (South Plainfield, New Jersey).

- Proteinase is purchased from Qiagen (Hilden, Germany).

- 5-Hydroxycytosine (5-OHCyt) is prepared using the method reported by Moshel and Behrman (1974).

- 5-OHdCyd and 5-OHdUrd are prepared by collidine treatment of 5-bromo-2'-deoxycytidine and 5-bromo-2'-deoxyuridine, respectively.

- The latter nucleosides are obtained by addition of bromine to a solution of 2'-deoxycytidine and 2'-deoxyuridine, respectively.

- Condensation of $[^{15}N_2]$-uracil with $[^2H_2]$-*para*formaldehyde (MSD Isotopes, Montreal) provided $[^{15}N_2,^2H_2]$-5-HMUra (Decarroz *et al.*, 1986).
- An aliquot fraction of the latter product is subsequently oxidized into $[^{15}N_2,^2H_2]$-5-ForUra (Brossmer and Ziegler, 1966).
- $[^{13}C,^{15}N_2]$-5-OHCyt is prepared as described for 5-OHCyt, using $[^{13}C,^{15}N_2]$-Cyt.
- The latter compound is synthesized from $[^{13}C,^{15}N_2]$-urea (Eurisotop, St Aubin, France) according to Bendich *et al.* (1949).
- $[^{15}N_2,^{18}O]$-5-OHUra is prepared from $[^{15}N_2]$-uracil by bromination followed by collidine treatment in $[^{18}O]$-H_2O (Eurisotop, St Aubin, France).
- $[^{13}C,^{15}N_3]$-8-oxoAde and $[^{13}C,^{15}N_3]$-8-oxoGua are synthesized according to Stadler *et al.* (1994).
- $[^{15}N_3]$-FapyGua and FapyGua are prepared from 2,4,6-trihydroxypyrimidine as published (Douki *et al.*, 1997).

DNA EXTRACTION FROM LIVER OR CELL CULTURES

Homogenization of the liver

1. Keep liver samples at $-80°C$ before use.
2. Weigh liver samples still frozen and divide into fractions of 150–250 mg (typically two or three per tube).
3. Homogenize each portion of tissue by using a 2 ml potter glass homogenizer in 1 ml of homogenization buffer (0.1 M NaCl, 30 mM TRIS, 10 mM EDTA, 5 mM desferroxamine mesylate, 0.5% Triton-100, 10 mM β-mercaptoethanol, pH 8) kept in ice.
4. After homogenization, transfer the sample into a 15 ml Falcon tube and centrifuge at 1300 $\times$ g for 15 min at 4°C.
5. Discard the supernatant and add 0.5 ml of homogenization buffer
6. Vortex the sample for 10 s and further centrifuge at 1300 $\times$ g for 3 min at 4°C. Then the supernatant is discarded.

Extraction of the nucleic acids

1. Suspend the nuclear pellet obtained from liver as previously described, or the cell pellet (5–10 $\times$ 10^6 cells) obtained by

centrifugation of cell cultures, in 3 ml of extraction buffer (20 mM NaCl, 20 mM TRIS, 20 mM EDTA, 5 mM desferroxamine mesylate, pH 8).

2. Add SDS (10% in water, 200 μl).
3. Homogenize the resulting suspension in a 5 ml potter glass homogenizer and then transfer into a 15 ml Falcon tube.
4. Add Proteinase (100 μl, 1 mg/ml) and gently vortex the resulting suspension for 10 s.
5. Maintain the sample for 1 h in a water bath set at 37°C.
6. Add chloroform (1 ml) and vigorously vortex the sample for 30 s prior to centrifuging at 3000 $\times$ g for 5 min at room temperature.
7. Collect the aqueous (upper) phase using a glass Pasteur pipette and transfer into a 15 ml Falcon tube.
8. Add successively a 4 M solution of NaCl (300 μl) and cold ethanol (7.5 ml, −20°C).
9. Gently shake the sample until complete precipitation of the nucleic acids.
10. Centrifuge the sample at 3000 $\times$ g for 10 min at room temperature.
11. Discard the supernatant.
12. Rinse the nucleic acids pellet by gently adding 500 μl of 70% ethanol.
13. Discard the liquid phase and leave the sample at room temperature for 5 min.

DNA isolation

1. Solubilize the nucleic acids pellet in 1 ml of RNase buffer (10 mM TRIS, 1 mM EDTA, 2.5 mM desferroxamine mesylate, pH 7.4).
2. Add RNaseA (100 μl, 1 mg/ml) and RNase T1 (10 μl, 1000 units/ml).
3. Gently vortex the sample and incubate at 37°C for 1 h.
4. Add 100 μl of a 4 M NaCl solution and 2.5 ml of cold ethanol (−20°C).
5. Gently shake the sample until complete precipitation of DNA.
6. Centrifuge the sample at 3000 $\times$ g for 10 min at room temperature.

7. Discard the supernatant and rinse the pellet by gently adding 500 μl of 70% ethanol.

8. Discard the liquid phase and leave the sample at room temperature for 5 min.

9. Solubilize the DNA pellet in 1 ml of deionized water.

10. Determine the concentration of the solution spectrophotometrically, assuming that 1 unit of absorbance at 260 nm corresponds to a concentration of 50 μg/ml of DNA.

DNA HYDROLYSIS

Formic acid treatment for the analysis of 5-OHCyt, 5-OHUra, 5-HMUra, 5-ForUra, 8-oxoGua and 8-oxoAde

1. Transfer the DNA solution (300 μg) into a 2 ml screw-cap vial.

2. Add a solution (30 μl) of either $[^{15}N_3, ^{13}C]$-8-oxoGua (10 μM), $[^{15}N_3]$-8-oxoAde (10 μM) or a mixture of $[^{15}N_2, ^{13}C]$-5-OHCyt, $[^{15}N_2, ^{18}O]$-5-OHUra, $[^2H_2, ^2N_2]$-5-HMUra, $[^2H, ^2N_2]$-5-ForUra (10 μM each).

3. Freez-dry the sample overnight.

4. Add formic acid (88%, 1 ml) and flush the resulting solution with a stream of nitrogen for 10 min.

5. Close the vial and then hold it at 140°C for 45 min in an aluminum heating block.

6. Remove, the vial from the heating block and allow to cool down to room temperature.

7. Remove formic acid under vacuum in a Speed-Vac.

8. Add water (500 μl) and evaporate the sample to dryness again.

9. Add water (500 μl) again and freeze-dry the resulting solution overnight.

HF/pyridine-mediated release of FapyGua from DNA

1. Transfer an aliquot fraction of the DNA solution (300 μg) into a 1.5 ml polypropylene vial.

2. Add a 10 μM solution of $[^{15}N_3]$-FapyGua (30 μl) and freeze-dry the resulting mixture overnight.

3. Add a 70% w/w solution of hydrogen fluoride in pyridine (100 μl).
4. Vortex the sample for 5 s and then place in a water bath held at 37°C.
5. Mix calcium carbonate (300 mg) and 1 ml of water under magnetic stirring in a 20 ml glass vial.
6. Add the HF/pyridine solution to the calcium carbonate suspension under vigorous stirring.
7. Leave the sample under stirring for 2 min and then allow the suspension to decant.
8. Check the pH of the liquid phase with pH paper. If the pH is lower than 7, stir the sample for an additional 1 min. Allow, the suspension to decant and check the pH again.
9. Carefully collect the liquid fraction with a pipetman and transfer into a 1.5 ml polypropylene vial.
10. Centrifuge the sample is centrifuged at 5000 × g for 5 min at room temperature.
11. Collect the supernatant and transfer into a clean 20 ml glass vial.
12. Add a [1:1] mixture of water and ethanol (1 ml) to the first solid fraction.
13. Stir the resulting suspension for 2 min and transfer into the polypropylene vial used previously.
14. Centrifuge at 5000 × g for 5 min at room temperature.
15. Collect the supernatant and mix to the first liquid phase in the 20 ml glass vial.
16. Freeze-dry the resulting solution overnight.

Homogenization of the liver

PREPURIFICATION

Following hydrolysis, freeze-dried residues are obtained. They are suspended in the HPLC buffers used for the purification of the products of interest:

- 100 μl of a 25 mM solution of ammonium formate for the HPLC purification of either FapyGua, 5-OHCyt, 5-OHUra, 5-HMUra, 5-ForUra or 8-oxodGuo
- 50 μl of water for the analysis of 8-oxoAde. Then 75 μl of methanol is slowly added. Ethyl acetate (375 μl) is subsequently added to the resulting mixture.
- 500 μl of PBS for the immunopurification of 8-oxodGuo.

5-OHCyt, 5-OHUra, 5-HMUra, 5-ForUra and FapyGua

Both the formic acid and the HF/pyridine hydrolyzed samples are injected on an HPLC apparatus (Gilson, Villiers le Bel, France) consisting of two model 306 pumps controlled by a model 305 pump. Samples are injected using a 231 XL autoinjector and fractions of interest are collected on an FC204 fraction collector. The elution is monitored with a 111B UV spectrometer set at 280 nm connected to a D2500 integrator (Hitachi, Tokyo, Japan). The system is equipped with an octadecylsilyl silica gel column (Interchrom HC18–25F, 250 × 4.6 mm i.d., particle size 5 μm; Interchim, Montluçon, France).

1. Start the elution with a 25 mM solution of ammonium formate in water for 10 min.
2. Add methanol linearly over 15 min until its proportion reaches 20%.
3. Keep the latter composition for 5 min. The flow rate is 3 ml/min.
4. Collect the two fractions eluting between 3 and 4 min (FapyGua), 4 and 5 min (5-OHCyt and 5-OHUra) and 5 and 8 min (5-HMUra and 5-ForUra), respectively.
5. Concentrate the two fractions under vacuum and freeze-dry.

8-OxoAde

The HPLC apparatus described above is used with a home-packed semi-preparative (250 × 7 mm i.d., particle size 10 μm) silica gel column (Macherey-Nagel, Düren, Germany). The isocratic eluent is a [75:16:2] v/v/v mixture of ethyl acetate, methanol and water. The fraction eluting between 5 and 7.5 is collected. Then the HPLC solvent is evaporated under vacuum. The resulting residue is dissolved in water (500 μl) and freeze-dried.

8-OxoGua

HPLC purification

The separation of 8-oxoGua is achieved using a Sulpelco LC18-DB 5 μm column (25 cm $\times$ 4.6 mm i.d.) under isocratic conditions. The flow rate is set at 1 ml/min with 2% methanol in 50 mM ammonium acetate buffer, pH 5.5 as the mobile phase. Nucleobases are detected by a UV spectrophotometer set at 260 nm. The formic acid DNA hydrolysate is injected on to the column. A 1.5 ml 8-oxoGua containing fraction is collected between 9 and 10.5 min, just after the elution of guanine. The fraction is concentrated under reduced pressure and then freeze-dried overnight.

Immunopurification

Preparation of immunoaffinity columns   The immunoaffinity matrix can be prepared either from monoclonal antibodies (Mabs) obtained from ascite fluid and partially purified by ammonium sulphate precipitation (40% v/v) or from commercially available Mabs (Pharmingen, San Diego, California). Swell 1 g of the CNBr-activated sepharose 4B in about 50 ml of 1 mM HCl (1 g of gel gives 3.5 ml final volume). Filtration is achieved using a sintered glass.

1. Wash the gel twice with 100 ml of 1 mM HCL.
2. Filtrate the gel and wash with 5 ml of coupling buffer (0.1 M NaHCO$_3$, 0.5 M NaCl, pH 8.3).
3. Rapidly transfer the gel into a 50 ml Erlen–Mayer containing the antibody as a [1:1] w/v solution in coupling buffer.
4. Coupling is performed using 2 mg protein per ml of gel.
5. Keep the solution under mild agitation (do not vortex) at room temperature for 2 h.
6. Stop the agitation so that the gel can be decanted.
7. Remove the supernatant and block unbound sites by addition (two bed volumes) of 0.1 M Tris-HCl buffer.
8. Restart vigorous stirring for 2 h at room temperature.
9. Using a sintered glass, wash the gel with at least five bed volumes of sodium acetate (0.1 M, pH 4.0) and NaCl (0.5 M), coupling buffer and PBS buffer, respectively.

The antibody-bound gel can be stored at 4°C as a suspension in PBS buffer. For a long storage, 0.02% NaN$_3$ has to be added to the PBS buffer. Immunoaffinity columns are prepared by packing Bio-Rad polyprep columns with 0.5 ml of the immobilized gel.

Purification by immunoaffinity chromatography   Formic acid hydrolyzed DNA samples are applied to the immunoaffinity column at room temperature in 0.5 ml of PBS. The column is washed under gravity with 2.5 ml of PBS buffer, followed by 2.5 ml of H_2O. The bound chemicals are then eluted with

---

Isolation of modified bases and nucleosides from urine

1. To 2 ml of urine, add 10 μl of 10 μM (100 pmol) of [$^{15}N_3$, ^{13}C]-8-*oxoGua* and [$^2H_2$, $^2N_2$]-5-HMUra.
2. Filter the urine solution through a Millex HV 0.45 μm filter (Waters Millipore) and concentrate to 1 ml under reduced pressure.
3. Inject the resulting solution on to a semi-preparative octadecylsilyl silica gel column (250 × 10 mm i.d., 5 μm particle size) (Interchim, Montluçon, France) equipped with an RP-18 guard column.
4. The purification is performed using the Gilson HPLC apparatus described above.
5. Maintain the mobile phase, consisting of 25 mM ammonium formate, with a flow rate of 2 ml/min, for 15 min.
6. Increase then, the concentration of acetonitrile in the mobile phase linearly to reach 8% after 45 min.
7. Collect fractions of 1 ml for 5-HMUra and 8-oxoGua (and their corresponding isotopically labelled internal standards) between 12–13 min and 21–22 min, respectively.
8. Freeze-dry the collected fractions.

2.5 ml of MeOH. In a subsequent step, methanol is removed under reduced pressure.

GCMS ANALYSIS

Silylation

Prepurification provides a dry residue, which is subsequently solubilized in 100 μl of water and transferred into 2 ml screw-cap GCMS autosampler vials. The samples are evaporated to dryness under vacuum. Then 100 μl of a [50:50] v/v mixture of acetonitrile and silylation reagent is added. The latter compound is either *N-tert*butyldimethylsilyl-*N*-methyl-trifluoroacetamide for 5-OHCyt, 5-OHUra, 5-HMUra, 5-ForUra, or *N-bis*(trimethylsilyl) trifluoroacetamide for FapyGua, 8-oxoGua and 8-oxoAde. Vials are held at 120°C for 20 min in an aluminum heating block.

GCMS analyses

GCMS analyses are carried out on a 5890 Series II gas chromatograph and an MSD 5972 mass detector (Hewlett-Packard, Les Ulis, France) used in the single ion monitoring mode. The system is equipped with a capillary column (0.25 mm, 15 m) coated with a 0.1 μm film of methylsiloxane substituted by 5% phenylsiloxane (HP5-trace, Hewlett-Packard). The constant flow rate is 0.8 ml/min. The injection (injection volume 1 μl) is performed in the splitless mode with the temperature of the injection port set at 250°C. The column is maintained at 130°C for 1 min. Then the temperature is increased linearly to 280°C at a rate of 10°C/min. The ions collected and the respective retention times of the six silylated modified bases are as follows.

- **FapyGua** 7.0 min; m/z = 457 (FapyGua + 4 TMS), 442 (FapyGua + 4 TMS – methyl), 460 ([^{15}N$_3$]-FapyGua + 4 TMS), 445 ([^{15}N$_3$]-FapyGua + 4 TMS – methyl)
- **5-OHUra** 7.9 min; m/z = 413 (5-OHUra + 3tBDMS – tButyl), 417 ([^{15}N$_2$, ^{18}O]-5-OHUra + 3tBDMS – tButyl)
- **5-OHCyt** 8.7 min; m/z = 412 (5-OHCyt + 3tBDMS – tButyl), 415 ([^{15}N$_2$, ^{13}C]-5-OHCyt + 3tBDMS – tButyl)
- **5-ForUra** 6.0 min; m/z = 311 (5-ForUra + 2 tBDMS – tButyl), 314 ([^{15}N$_2$,^{2}H]-5-ForUra + 2 tBDMS – tButyl)
- **5-HMUra** 8.4 min; m/z 427 (5-HMUra + 3 tBDMS – tButyl), 431 ([^{15}N$_2$,^{2}H$_2$]-HMUra + 3 tBDMS – tButyl)
- **8-oxoAde** 5.6 min; m/z = 367 (8-oxoAde + 4 TMS), 352 (8-oxoAde + 4 TMS – methyl), 371 ([^{15}N$_3$,^{13}C]-8-oxoAde + 4 TMS), 356 ([^{15}N$_3$,^{13}C] – 8-oxoAde + 4 TMS – methyl)
- **8-oxoGua** 7.2 min; m/z = 455 (8-oxoGua + 4 TMS), 440 (8-oxoGua + 4 TMS – methyl), 459 ([^{15}N$_3$,^{13}C]-8-oxoGua + 4 TMS), 445 ([^{15}N$_3$,^{13}C]-8-oxoGua + 4 TMS – methyl).

Calibration curve

For each compound, 200 pmol of isotopically labelled internal standard in aqueous solution is added to a series of 6 GCMS vials. The compound of interest is added to reach final concentrations of 0, 10, 50, 100, 200 and 500 pmol, respectively. The samples are evaporated to dryness under vacuum prior to being silylated and analyzed by GCMS.

Interpretation of the results

For each analysis, the chromatograms corresponding to the different ions collected are traced separately. The peaks corresponding to the retention time of the products of interest (difference in the retention time: less than 0.02 min with respect to the standard) are integrated. For a given compound, the area of the peak corresponding to the normal molecule (or the sum of the two peaks when two ions are monitored, such as [M] and [M-15] for trimethylsilyl derivatives) is divided by the sum of the peak corresponding to the related isotopically labelled internal standard (or the sum of the two corresponding peaks). For the calibration curve, the ratio obtained is plotted against the concentration of the compound of interest in each of the six vials. A linear fitting is performed, providing a value for the slope. The amount (a, in pmol) of product present in the unknown samples is calculated by dividing the ratio between the ions of the normal and the labelled molecule by the value of the slope of the calibration curve. The amount of DNA (m, in μg) for each sample is obtained from the spectrophotometric determination of the concentration of the initial DNA solution. Considering that the average molecular weight of a single nucleotide is 324.5, the level of modification (in lesions per 10^6 normal bases) is $a/(w/324.5)$. A value of 1 lesion per 10^6 normal bases corresponds to 0.4 lesions per 10^5 base precursor or 0.0031 pmol/μg DNA. The latter calculations are based on an equal molar distribution of A:T and G:C base pairs in DNA. More precise results can be obtained by using the values determined for each species (for instance: calf 43.8% G:C pairs).

REFERENCES

Bendich, A., Gelter, H. and Brown, G.B. (1949) Synthesis of isotopic cytosine and a study of its metabolism in rat. *J. Biol. Chem.*, **177**.

Bianchini, F., Hall, H., Donato, F. and Cadet, J. (1996) Monitoring urinary excretion of 5-hydroxymethyluracil for assessment of oxidative DNA damage and repair. *Biomarkers*, **1**, 178–184.

Brossmer, R. and Ziegler, D. (1966) Zür darstellung heterocyclisher aldehyde. *Tetrahedron Lett*, **43**.

Decarroz, C., Wagner, J.R., van Lier, J.E., Murali, K.C., Riesz, P. and Cadet, J. (1986) Sensitized photo-oxidation of thymidine by 2-methyl-1,4-naphtoquinone. Characterization of stable products. *Int. J. Radiat. Biol.*, **50**.

Douki, T., Delatour, T., Bianchini, F. and Cadet, J. (1996a) Observation and prevention of an artefactual formation of oxidized DNA bases and nucleosides in the GC–EIMS method. *Carcinogenesis*, **17**, 347–352.

Douki, T., Delatour, T., Paganon, F. and Cadet, J. (1996b), Measurement of oxidative damage at pyrimidine bases in γ-irradiated DNA. *Chem. Res. Toxicol.*, **9**, 1145–1151.

Douki, T., Martini, R., Ravanat, J.-L., Turesky, R.J. and Cadet, J. (1997) Measurement of 2,6–diamino-4-hydroxy-5 formamidopyrimidine and 8–oxo-7,8–dihydroguanine in isolated DNA exposed to gamma radiation in aqueous solution. *Carcinogenesis*, **18**, 2385–2391.

Moshel, R.C. and Behrman, E.J. (1974) Oxidation of nucleic acid bases by potassium peroxodisulfate in alkaline aqueous solution. *J. Org. Chem*, **39**.

Ravanat, J. L., Turesky, R.J., Gremaud, E., Trudel, L.J. and Stadler, R.H. (1995) Determination of 8oxoguanine in DNA by gas chromatography–mass spectrometry and HPLC–electrochemical detection: overestimation of the background level of the oxidized base by the gas chromatography–mass spectrometry assay. *Chem. Res. Toxicol.*, **8**, 1039–1045.

Stadler, R.H., Staempfli, A.A., Fay, L.B., Turesky, R.J. and Welti, D.H. (1994) Synthesis of multiply-labeled [$^{15}N_3$,$^{13}C_1$]-8-oxo-subsituted purine bases and their corresponding 2′-deoxynucleosides. *Chem. Res. Toxicol.*, **7**, 784–791.

3 The Measurement of Protein Oxidation by HPLC

Helen R. Griffiths, Ruth Bevan and Joe Lunec

INTRODUCTION

The complexity of protein structure, due not only to the 21 amino acids that are involved in the protein backbone, but also to the involvement of carbo-hydrates in stabilizing structures, has hindered advances in the analysis of protein oxidation. Yet the post-synthetic modification of proteins by *in vivo* oxidation can dramatically alter their biological activity. This is exemplified by oxidative modifications to histidine and lysine in apolipoprotein B present in LDL (Chapter 1) (Steinberg *et al.*, 1989; Esterbauer *et al.*, 1992), which alters receptor recognition of the particle. Altered immunological activity of IgG has also been demonstrated following exposure to specific radical species, where the effect on biological activity depends on the nature of the denaturing species (Griffiths *et al.*, 1988). This invokes the concept of amino acid-specific oxidative damage from distinct radical species, and highlights the importance of improved detection methods for oxidatively modified amino acids within proteins.

Over the last 10 years, the measurement of protein oxidation has increased in popularity with the advent of sensitive and specific measures of amino acid and protein oxidation by HPLC.

This chapter will describe HPLC techniques and point out pitfalls to enable the reader to establish valid and appropriate methods for the measurement of protein oxidation. The first section will consider methods that look at the whole protein, whilst the second half will consider methods for amino acids from protein hydrolysates. Within this framework, sample preparation will be considered, including protein hydrolysis.

Measuring in vivo *Oxidative Damage: A Practical Approach.* Edited by J. Lunec and H. R. Griffiths. © 2000 by John Wiley & Sons, Ltd. ISBN 0 471 81848 8.

MARKERS OF PROTEIN OXIDATION IN INTACT PROTEINS

Induction of novel fluorescence

All proteins exhibit a native UV fluorescence by virtue of the intrinsic aromatic amino acids, particularly tryptophan and tyrosine. When these amino acids become modified, their fluorescence spectrum is altered with characteristic excitation (Ex.) and emission (Em.) spectra lying within the visible region. This phenomenon can be harnessed by applying postcolumn fluorimetric detection to protein eluates separated on a gel column by HPLC.

Equipment and reagents
- Phosphate buffer, pH 7.2 (9.1 g/l KH_2PO_4, 7.5 g/l KCl)
- TSK 3000SW gel column, 25 cm × 9.2 mm
- TSK 3000SW guard column, 10 cm × 9.2 mm
- Isocratic HPLC system fitted with rheodyne injector valve and 20 μl loop. UV (280 nm) and fluorimetric detection (Ex. 360 nm, Em. 454 nm)
- Centrifuge
- Whole blood or proteins under study
- HPLC grade water
- (Amicon microcentrifuge)

Methods

Sample preparation
1. Collect blood into anticoagulant-free tubes and allow to clot at room temperature.
2. Centrifuge the clotted blood at 100 × g for 10 min at 4°C to prepare the serum. Sera can be stored at −20°C for up to a month if necessary.

HPLC
1. Prepare a standard solution of proteins at 1 mg/ml final concentration in PBS; albumin, catalase and IgG are recommended as these are well separated on the TSK column. In addition, the first two proteins exhibit significant visible

fluorescence. These are stable at 4°C for a week and should be allowed to equilibrate to room temperature before analysis.

2. Prepare HPLC eluent as follows: 9.1 g/l K$_2$HPO4: 7.5 g/l KCl, adjusted to pH 7.2 with KOH.

3. Inject 50 μl of standard or sample via the rheodyne injector valve. Elute at a flow rate of 1 ml/min and monitor with a UV detector set at 240 nm and 0.05 AUFS, and fluorescence detection set at Ex. 360 nm and Em. 454 nm, set at 100 × gain.

4. Representative chromatogram is shown in Figure 3.1.

Data handling

1. Internal standardization may be achieved by the addition of fluorescein labelled-cytochrome *c* to the samples under investigation, since its low molecular weight causes it to be retained on the column after other serum proteins have been eluted.

2. By injecting small sample volumes, the column efficiency is maintained; however, regular column clean-up is recommended, according to manufacturer's instructions, when the column is used heavily for serum samples. The use of external standards at the start of the run will confirm retention times and enable column condition to be evaluated.

Application of the assay to biological samples

The fluorescence intensity of IgG has been the most frequently cited, since the high fluorescence of albumin can be attributed to both its lipid and drug-carrying capacity. In addition, the glycation of albumin is commonplace, and the ensuing oxidative reactions can further enhance protein fluorescence.

Using a measure of IgG fluorescence, an increase in oxidative activity has been demonstrated in ischaemia/reperfusion, rheumatoid arthritis and diabetes (Lunec *et al.*, 1985; Jones and Lunec, 1987; Ward *et al.*, 1994). This methodology represents a simple procedure for determining oxidative damage to proteins *in vivo* or *in vitro*, but it cannot give any information on the potential for modified function. A more detailed evaluation of amino acid or peptide oxidation is required for such prediction.

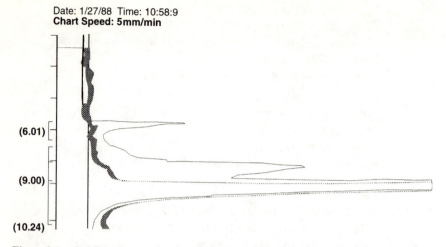

Date: 1/27/88 Time: 10:58:9
Chart Speed: 5mm/min

(6.01)

(9.00)

(10.24)

Figure 3.1   Gel filtration chromatogram of a serum sample (50 μl), monitored by UV absorbance at 280 nm (———) and fluorescence (--------, Ex. 360 nm, Em. 454 nm), where the peak eluting at 6.01 min corresponds to aggregates and IgM, the peak at 9.00, corresponds to IgG, and where the peak at 10.24 min is albumin.

Carbonyl measurement

Protein carbonyls are formed either as a result of degradative changes to the protein backbone (for example, during proline degradation and chain break-age) or through association of carbonyl products derived from other oxidat-ive processes, such as lipid peroxidation. This is common in low density lipoprotein, where the protein component (apo B) is extensively modified by malondialdehyde (MDA) and hydroxy-nonenal (HNE) (Young and Trim-ble, 1991; Kinter et al., 1994), and also in membrane-bound proteins such as sarcoplasmic reticulum Ca^{2+} ATPase (Viner et al., 1997). The determination of total protein carbonyls present in a biological sample has been widely applied, using their specific reaction with 2,4-dinitrophenyl hydrazine (DNPH). Following Folch extraction and removal of lipids, it is dependent on the formation of a carbonyl-hydrazone from hydrazine under acidic conditions, which can be detected spectrophotometrically at 370 nm.

This gives little information as to the relative susceptibility of different proteins to carbonyl formation, is subject to interference in complex mixtures and requires a large sample size for spectrophotometric determinations. Therefore, the HPLC separation of DNPH-derivatized samples has been adopted, enabling the proportion of carbonyls present in distinct proteins to be determined (Jones et al., 1956). The majority of papers cite the use of a method utilizing guanidine-HCl; however, this causes a lot of back pressure and can cause corrosion of filters etc. Therefore, Levine *et al.* (1994) have described a less aggressive method, using SDS in the mobile phase, where the effects of salt corrosion are virtually eliminated.

Equipment and reagents
- Isocratic HPLC system fitted with rheodyne injector valve and 50 μl loop, with dual UV wavelength detection fixed at 276 and 360 nm.
- 200 mM sodium phosphate buffer, pH 6.5, containing 1% SDS
- Zorbax GF 450, 92 × 250 mm column, and 92 × 100 mm guard column
- Biological sample (up to 10 mg protein/ml is recommended– dilute in PBS if necessary)
- 12% SDS
- 20 mM DNPH
- 10% Trifluoracetic acid (TFA)
- 2 M Tris base solution
- HPLC water

Methods

Sample preparation
1. Prepare a solution of 10% (v/v) TFA in water.
2. When calculating the amount of DNPH required for a 20 mM solution, take into account the water content of the DNPH, which should be stipulated by the manufacturer and may be as great as 30%.
3. To facilitate dissolution of DNPH in TFA, gentle warming or sonication is recommended.
4. Precipitates that appear during storage can be removed by gentle centrifugation.
5. To derivatize carbonyls, take 100 μl of sample and add 100 μl 12% SDS. At least 6% SDS is required for derivatization.
6. To the SDS-treated sample, add 200 μl DNPH in TFA and vortex to mix.
7. A derivatization blank should be prepared by treating the SDS sample with an equal volume of 10%TFA only.

8. Reaction should be allowed to proceed at room temperature for 10 min.
9. Terminate the reaction by adjusting the pH to ~ 7, using 2 M Tris. This will cause a colour change from yellow to red/orange. It is important to keep the pH below 8 to avoid alkali-induced changes in the hydrazones.

HPLC
1. Inject 50 μl of sample on to Zorbax column through rheodyne injector.
2. Elute derivatized sample and sample blanks at a flow rate of 2 ml sodium phosphate buffer/min.
3. Monitor eluent at 276 nm set at 0.5 AUFS, and at 360 nm for the formation of hydrazones.
4. A representative chromatogram is shown in Figure 3.2.

Data handling
1. The carbonyl content of samples can be referred to as nmol/mg protein or as mol carbonyl/mol protein. For either measurement, it is necessary to calculate the absolute amount of carbonyl by integrating the chromatograms. Since the excess hydrazine elutes at the end of the run, it is advisable to stop integrating prior to its elution.
2. The absorption coefficient for the hydrazone is 22 000 at 360 nm, and 9460 at 276 nm.
3. The molar absorption coefficient of the protein of interest should also be known, but 50 000 can be substituted as an estimate for an average protein if unknown.
4. The following calculation should be applied:

$$\text{mol carbonyl/mol protein} = \frac{\text{E (protein)} \times \text{area (hydrazone)}}{\text{E (hydrazone)} \times \text{area (\textit{protein})}}$$

where E (hydrazone) = 22 000; E (protein), if unknown, can be substituted with 50 000; area (hydrazone) = area calculated under the peak at 360 nm; and area (protein) = area calculated under the peak at 270 nm less the area at 270 nm contributed by hydrazone $\equiv 0.4 \times$ peak area at 360 nm.

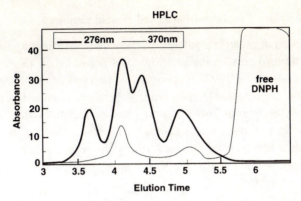

Figure 3.2   Gel fitration chromatogram of mixed protein solution, monitored by UV (dark line) (276 nm) and visible (fine line) (370 nm) absorbance, where protein-bound hydrazones are 4.1 and 5.00 min, and unbound DNPH elutes after 5.5 min. (From Schacter *et al.*, 1996.)

Applications

This methodology is far more robust than the more frequently cited guanidine-HCl method and is less corrosive on the HPLC. Perhaps the only slight restriction on its application is that peak resolution is less well defined under SDS, but for well separated proteins this should not be a problem. For detailed methodology of the guanidine-HCL technique the reader is referred to Shacter *et al.*, 1996.

The development of immunological reagents for the detection of carbonyls has facilitated the detection of carbonyls in proteins separated by SDS-PAGE; however, this is at most a qualitative assessment.

OXIDIZED AMINO ACID ANALYSIS

In order to undertake a more detailed analysis of protein oxidation, and in particular to study susceptibility of distinct amino acids to oxidation *in vivo*, it is necessary to undertake protein isolation and hydrolysis prior to analysis. The greater the manipulations that oxidized proteins are exposed to, the greater is the likelihood of further oxidation. Precautions should therefore be taken in order to minimize *in vitro* oxidation and these will be alluded to below.

The following sections describe methods available for isolation of proteins of interest, and their subsequent hydrolysis, prior to any consideration of the oxidized amino acid analysis itself.

Isolation of oxidized proteins from mixed protein samples

There has been considerable interest in the role of oxidized apolipoproteins carried by lipoproteins, in the aetiology of atherosclerosis (Steinberg *et al.*, 1989; Esterbauer *et al.*, 1992). It is postulated that oxidation of critical amino acids required for receptor binding renders them unrecognised by their native receptor, but as ligands for a second class of receptors that accumulate lipoproteins in a non-regulated manner. Therefore, study of the oxidation of apoproteins has received intense scrutiny. In order to carry out oxidized amino acid analysis in the absence of contaminating lipid, an extraction must first be undertaken.

Apolipoprotein extraction

Equipment and reagents
- Chloroform
- Methanol
- Lipoproteins isolated from plasma by differential gradient ultracentrifugation
- Vortex
- Centrifuge
- Desiccator

Method
1. To 3 ml of LDL, add 2 ml ice-cold chloroform:methanol (2:1).
2. Vortex, and centrifuge for $3000 \times g$ for 10 min at $10°C$.
3. Remove the chloroform layer, and discard.
4. Repeat this three times, to ensure complete washing of the protein.
5. Most of the precipitate remains as a smeary film on the walls of the tube.
6. Evaporate to dryness and freeze-dry the protein for hydrolysis.

Notes
Ether/acetone extraction of Lpa is also reported (Kikugawa *et al.*, 1991). The use of immobilized antibodies is one potential route to allow a wide variety of proteins to be isolated, where antibodies specific to the protein of interest are linked to sepharose, or

magnetic beads. The mixed protein solution is then passed through the column, where the protein under study is retained. The protein can be eluted subsequently, using a general dissociating agent. Refer to (Johnstone and Thorpe 1987) for further information.

Protein hydrolysis

Prior to amino acid analysis, the protein must be hydrolysed into its constituent amino acids. There are two common methods of hydrolysis: (i) acid hydrolysis; and (ii) enzymic hydrolysis. These will be described in turn. Whilst enzymic hydrolysis appears to be the most widely applicable, the protein under study and the products of interest should be studied prior to determining the method of choice.

Acid hydrolysis

Equipment and reagents
- Protein under study, at a minimum of 1 mg/ml.
- desiccator/freeze-drier
- 6 M HCl
- 0.01 M phenol
- Boiling bath/incubator at 105–110°C

Method
1. Freeze-dry 1 ml of the sample in analysis vessel.
2. Add 6 M HCl containing 0.01 M phenol and thoroughly deoxygenate using a stream of nitrogen.
3. Heat to 105–110°C for 24–27 h.
4. Lyophilize the sample and store at −70°C for subsequent HPLC analysis. The reconstitution buffer depends on the analytical method.

Notes
The use of phenol in this method is cautioned, since we have previously demonstrated that extraction of oxidized DNA using

phenol accentuates the degree of oxidative damage (Finnegan *et al.*, 1996).

Tryptophan is very susceptible to destruction by HCl hydrolysis, particularly in the presence of oxygen, heavy metals or carbohydrates. However, good yields can be obtained by hydrolysis with other acids, such as 3 M mercaptoethanesulphonic acid (Penke *et al.*, 1974). Unfortunately, hydrolysis is not complete until 96 h. In addition, hydroperoxides decompose to hydroxides under acidic hydrolysis.

The preferred procedure for degradation of proteins is using enzymic hydrolysis.

Enzymic hydrolysis

Equipment and reagents
- Protein under study
- Pronase E or proteinase K
- PBS
- Penicillin/streptomycin
- 50 mM mannitol
- Freeze-drier

Methods
1. To 2.9 ml of protein solution in PBS (10 mg protein/ml), add 0.1 ml of pronase E (3.67 mg/ml).
2. Add mannitol, to a final concentration of 50 mM, to act as an antioxidant.
3. Add streptomycin/penicillin to act as a bacteriostat.
4. Incubate at 37°C in the dark for 24 h.
5. Freeze-dry the hydrolysate until analysis.
6. Blanks containing all of the above cocktail, except the protein under study, should be run for each hydrolysis.
7. The pronase is itself hydrolysed and, being only 1% by weight of the total mixture, does not interfere with subsequent analysis. The blank will confirm this. A standard protein should

be hydrolysed with each batch to evaluate the efficiency of
hydrolysis.

Having completed hydrolysis of the protein, there are several different
analytical procedures available for the study of amino acid oxidation. To
get a complete picture it is necessary to carry out all the procedures, since
each method detects different products of protein oxidation. More realisti-
cally, the method of choice depends on the protein under study, and the free
radical denaturing system employed. For example: hydroperoxide yields are
high following gamma radiolysis; tryptophan oxidation is prevalent after UV
exposure; nitrotyrosine is formed from the effects of NO on tyrosine; and
carbonyls are common in metal-catalysed reactions.

The remainder of this chapter will describe four methods for the detection
of tryptophan oxidation, tyrosine oxidation, nitrotyrosine formation and
valine hydroperoxides.

Tryptophan oxidation

The oxidation of aromatic amino acids is associated with the loss of native
fluorescence in the UV region and the induction of a novel autofluorescence
in the visible region. For tryptophan, the oxidation through hydroperoxide
intermediates to yield three major products – HPI, kynurenine and N-formyl
kynurenine (NFK) – is reported. Of these, NFK is formed in greatest yield,
and can be easily detected because of its intrinsic fluorescence (Griffiths *et
al.*, 1992).

Equipment and reagents
- Cobalt 60 gamma source
- 1 mM tryptophan
- 1 mM tyrosine
- 1 mM phenylalanine
- 1 mM 5HT
- Amino acid hydrolysate
- Acetonitrile

- 0.15 M phosphate buffer, pH 5.2
- Gradient HPLC system with controller, injector valve and 50 μl loop. UV detector at 260 nm, fluorescence excitation at 360 nm and emission at 454 nm
- C-18 5 μm reversed phase column 25 cm × 4.6 mm i.d.) and guard column (10 cm × 4.6 mm i.d.) with same packing material

Methods

Sample preparation

1. Standards should be prepared by exposure of tryptophan in 40 mM phosphate buffer to 1000 Gy from a cobalt 60 gamma source. For NFK production, see Griffiths *et al.* (1992).
2. Protein hydrolysates can be injected after resuspending in a fixed volume of HPLC-grade water, from pronase E hydrolysed samples.

HPLC

1. Prepare a 50% solution of acetonitrile in phosphate buffer. The use of absolute acetonitrile can affect the performance of pump seals. Establish the following gradient profile for measurement of NFK:

Time (min)	0	2	4	6	8	17	18
Acetonitrile (50%)	2	4	8	14	20	20	2
Phosphate buffer	98	96	92	86	80	80	98

2. Allow the column 10 min to re-equilibrate at the end of each run prior to injection of further samples.
3. Inject 20–50 μl of 1 mM irradiated tryptophan solution, and elute at a flow rate of 1 ml/min according to the gradient tabulated above.
4. The identity of each product can be confirmed from a knowledge of its UV absorption spectrum, and by mass spectrometry.

Data handling

A standard curve for tryptophan, *N*-formylkynurenine and kynurenine can be established from which absolute levels of oxidized amino acids within proteins can be calculated.

A standard chromatogram is illustrated in Figure 3.3 showing the retention of tyrosine, HPI, K and NFK. Figure 3.4 shows a standard chromatogram of the hydrolysate of irradiated BSA.

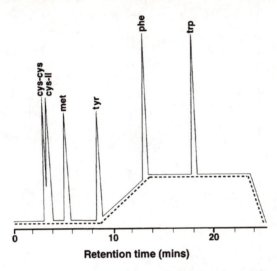

Figure 3.3 Reverse-phase chromatogram of native amino acids, monitored at UV absorption (240 nm- - -) and visible fluorescence (Ex. 360 nm, Em. 454 nm ····).

Applications

For studying albumin, this method is relatively insensitive, since BSA only contains one tryptophan residue/molecule. However, in IgG, tryptophan oxidation within proteins can be detected after a dose of 50 Gy. Using this and related methodology, increased levels of NFK have been reported at raised levels in IgG from patients with the chronic inflammatory disease, rheumatoid arthritis (Griffiths *et al.*, 1988).Recently, NFK has been identified in the apolipoprotein B component of freshly isolated LDL, and levels were shown to increase in LDL during *in vitro* oxidation (Giessauf *et al.*, 1996).

Tyrosine oxidation

Under certain oxidizing conditions, tyrosine can cross-link to form bityrosine, which, like the tryptophan oxidation products, has a visible

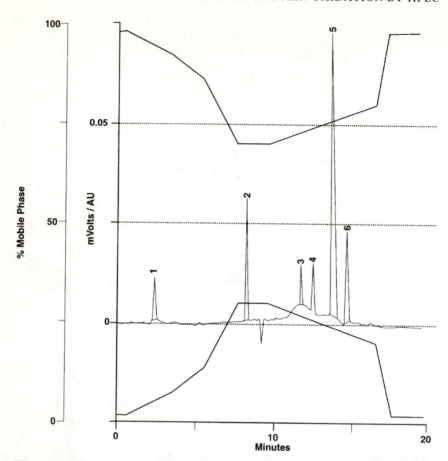

Figure 3.4 Reverse-phase chromatogram of pronase hydrolysed gamma-irradiated bovine serum albumin, monitored by UV absorption (240 nm- - -) and visible fluorescence (Ex. 360 nm, Em. 454 nm ····). Peak 1, cysteine; Peak 2, tyrosine; Peak 3, phenylalanine; Peak 4, kynurenine; Peak 5, tryptophan; Peak 6. *N*-formyl kynurenine.

autofluorescence (Ex. max. 280 nm, Em. max. 410 nm). Based on a knowledge of the chemistry involved in bityrosine formation, which proceeds via a phenoxyl radical intermediate, in general under aerobic conditions, this intermediate should be 'repaired', and bityrosine will not be formed. However, in the case where the intermediate phenoxyl radical can be stabilized by haem (for example, in sperm whale myoglobin), bityrosine is formed (Tew *et al.*, 1988). Some workers have implied from simple visible fluorescence measurements (Ex. 330 nm, Em. 430 nm), and the presence of non-reducible cross-links, that bityrosine is formed. This is a gross oversimplification, and is addressed at length by O'Connell *et al.* (1994). Bityrosine can only be confirmed by HPLC analysis. Several methods are reported based on reverse-phase HPLC with acidified organic solvent elution of bityrosine and fluorescence detection (Daneshvar *et al.*, 1997; Giulivi and Davies, 1994; O'Connell *et al.*, 1994).

Equipment and reagents
- Gradient HPLC system with controller, UV (276 nm) and fluorescence (Ex. 330 nm, Em. 410 nm), fitted with 50 μl loop.
- C-18 supelco 5 μM 25 × 1 cm (4.6 mm i.d.) with C-8 guard column (O'Connell *et al.*, 1994)
- Methanol
- Trifluoracetic acid
- Sodium hydroxide
- Tyrosine

Methods

Sample preparation
1. Prepare a standard solution of bityrosine. There are several reported methods, which differ in yield and complexity. By far the simplest is from the irradiation of a 1 mM solution of tyrosine in 40 mM phosphate buffer (pH 7.4) which has been deoxygenated by bubbling through with nitrogen. Column purification of the fluorescent bityrosine from the native tyrosine should be undertaken.
2. Resuspend lyophilized sample in 0.1% TFA.

HPLC
1. Establish a linear gradient of increasing methanol in water/0.1% TFA (adjusted to pH 2.5 with NaOH), from 15–20%.
2. Inject 50 μl of standard or protein hydrolysate.
3. Elute at a flow rate of 0.8 ml/min and monitor the eluent with a fluorescence detector set at Ex. 280 nm and Em. 410 nm.

Data handling
The concentration of the bityrosine standard preparation is quantified against the fluorescence of diphenol, where the detection limit for bityrosine is 1 pmol.

Notes
Bityrosine standard can also be prepared from treatment of 10 mM leu-enkephalin for 1 h at 37°C with AAPH (0.5 mM) in 10 mM KPO$_4$, 100 mM NaCl (Guilivi and Davies, 1994). Other workers have generated bityrosine from peroxidase/hydrogen peroxide treatment (Daneshvar *et al.*, 1997).

O'Connell *et al.* (1994) have shown that even if all the bityrosine formed from hypochloric acid treatment of Lp[a] was due to interchain cross-linking, this could not account for the total cross-linking. This indicates that studies looking at cross-linking and fluorescence without HPLC analysis are prone to misinterpretation.

Nitrotyrosine formation

Peroxynitrite is a powerful oxidant that has been implicated in macrophage-mediated bacterial cell killing, in which it exerts its toxicity through sulphydryl oxidation. Furthermore, protonated peroxynitrite can decompose by homolytic fission to produce the hydroxyl radical and nitrogen dioxide, resulting in nitration and hydroxylation of aromatic rings on amino acids such as tyrosine within proteins. (Hibbs *et al.*, 1988)

Equipment and Reagents
- 3-Nitro-L-tyrosine (Aldrich)
- Isocratic HPLC with rheodyne injector valve (50 μl loop) and UV detector set at 275 nm
- C-18 RP column, 4.6 × 250 mm, with a guard column
- 0.03 M citrate, 0.03 M acetate buffer (pH 3.6)

Methods

Sample preparation
Authentic standard for 3-nitrotyrosine should be used to generate
a calibration curve.

HPLC
1. Inject 50 μl of sample on to the column.
2. Analysis of standards and samples is achieved by isocratic
   elution with 0.3 M citrate/acetate buffer at a flow rate of
   0.7 ml/min, and monitoring the eluent at 275 nm.
3. Under these conditions, nitrotyrosine elutes at 12 min.

Figure 3.5 shows a standard chromatogram.

Data handling
The amount of nitrotyrosine in unknown samples can be calculated using peak areas of known concentrations of standard nitrotyrosine solutions, to establish a standard curve.

Applications

Nitrotyrosine has been detected in synovial fluids from patients with
rheumatoid arthritis (Kaur and Halliwell, 1994). Using animal models of
inflammation, increased plasma levels of nitrotyrosine have been detected
(Evans *et al.*, 1996). However, in another study, atherosclerotic lesions were
not found to contain detectable levels of nitrotyrosine (Shigenaga *et al.*, 1997).

Valine hydroperoxide analysis

One of the longer-lived reactive species produced by oxygen radical attack of
proteins is the hydroperoxide species. In time, these oxidizing species decay
to hydroxides, both of which can be measured by HPLC. The most susceptible amino acid to hydroperoxide formation is valine; therefore measurement
of valine hydroperoxide will afford greatest sensitivity (Gebicki and Gebicki,
1993). Fu *et al.* (1995) have described the identification and detection of these
species.

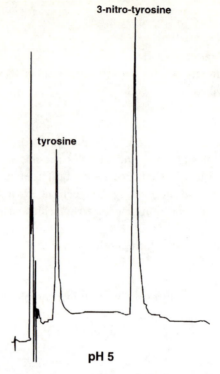

Figure 3.5 Reverse-phase chromatogram of 3-nitrotyrosine and nitrotyrosine prepared in pH 5 buffer. Eluent was monitored at 270 nm, and nitrotyrosine eluted at 12 min.

Equipment and reagents
- HPLC water
- Valine
- Isoluminol
- Microperoxidase
- Cobalt 60 gamma source
- Catalase
- Isocratic HPLC system with UV detector (210 nm) and chemiluminescence detector
- 20 μl injection port
- Microperoxidase
- Methanol

- Acetonitrile
- 10 mM phosphate buffer (pH 4.3)
- LC-NH$_2$, 5 μM particle size column (25 cm × 4.6 mm) with guard column
- Sodium borate buffer (0.1 M, pH 10)
- post-column pump to deliver chemiluminescence reagent
- T-piece for mixing

Method

Sample preparation
1. Irradiate valine for a dose of 1200 Gy while gassing continuously with oxygen.
2. Add 5 μg/ml catalase to the amino acid, post irradiation, to remove any hydrogen peroxide present.

HPLC
1. Inject 20 μl of sample on to column and elute with 80% acetonitrile in 10 mM phosphate buffer (pH 4.3) at a flow rate of 1 ml/min.
2. Monitor eluent with a UV detector set at 210 nm.
3. Post-UV detection, mix the eluent with chemiluminescence reagent, consisting of 1 mM isoluminol and 6 mg microperoxidase in 41 of sodium borate buffer (0.1 M, pH 10) and methanol [1:1].
4. Reagent should be delivered at a flow rate of 1.5 ml/min with the eluent and allowed to mix in the T-piece before passing to the chemiluminescence detector.

Data handling
A standard curve can be established using *tert*butyl hydroperoxide, in order to calculate absolute hydroperoxide levels.
Figure 3.6 shows a standard chromatogram.

Applications
This methodology will give information as to recent oxidative events, as there is a potential for repair of such lesions by glutathione peroxidase. In addition, the short half-life of these species enables recent events to be

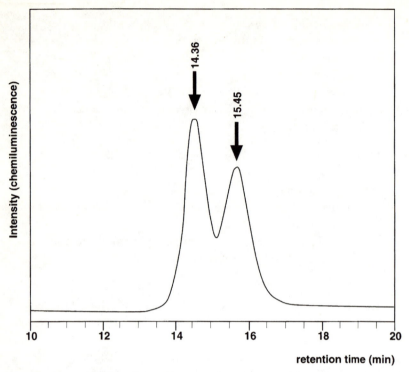

Figure 3.6 Post-column chemiluminescence detection of valine hydroperoxides separated on an LC-NH$_2$ column (From Fu *et al.*, 1995.)

followed. As yet, there is little available data in the literature as to the physiological levels of hydroperoxides in biological fluids. However, as described earlier, hydroperoxides are reduced during acid hydrolysis, and the measurement of OPA-derivatized hydroxides is recommended.

CONCLUSION

There has been an exponential rise in the number of publications relating to oxidative damage to proteins over the last 15 years, and this is largely due to increasingly sensitive methods of detection. There are still many unanswered questions relating to the stability of oxidized proteins and the effects on their functions. This is complicated by the lack of available quality assurance material, particularly in the commonly used carbonyl assay. The post-synthetic oxidation of proteins is likely to make an important contribution to both pathological and toxicological processes, where modification to protein function can initiate a downstream cascade of intracellular effects.

ACKNOWLEDGEMENTS

The authors acknowledge financial support from the Ministry of Agriculture, Fisheries and Food for their evaluation of methods for studying protein oxidation.

REFERENCES

Daneshvar, B., Frandsen, H., Dragsted, L.O. *et al.* (1997) Analysis of native human plasma proteins and haemoglobin for the presence of bityrosine by high-performance liquid chromatography. *Pharmacol. Toxicol.*, **81**, 205–208.

Esterbauer, H., Gebicki, J., Puhl, H. and Jurgens, G. (1992) The role of lipid-peroxidation and antioxidants in oxidative modification of ldl. *Free Radical Biol. Med.*, **13**, 341–390.

Evans, P., Kaur, H., Mitchinson, M.J. and Halliwell, B. (1996) Do human atherosclerotic lesions contain nitrotyrosine? *Biochem. Biophys. Res. Comms*, **226**, 346–351.

Finnegan, M.T., Herbert, K.E., Evans, M.D., Griffiths, H.R. and Lunec, J. (1996) Evidence for sensitisation of DNA to oxidative damage during isolation. *Free Radical Biol. Med.*, **20**, 93–98.

Fu, S., Hick, L.A., Sheil, M.M. and Dean, R.T. (1995) Structural identification of valine hydroperoxides and hydroxides on radical-damaged amino-acid, peptide, and protein molecules. *Free Radical Biol. Med.*, **19**, 281–292.

Gebicki, S. and Gebicki, J.M. (1993) Formation of peroxides in amino-acids and proteins exposed to oxygen free radicals. *Biochem. J.*, **289**, 743–749.

Giessauf, A. van, Wickern, B., Simat, T. *et al.* (1996) Formation of N-formylkynurenine suggests the involvement of apolipoprotein B-100 centered tryptophan radicals in the initiation of LDL lipid peroxidation. *FEBS Lett*, **389**, 136–140.

Giulivi, C. and Davies, K.J.A. (1994) Dityrosine – a marker for oxidatively modified proteins and selective proteolysis. *Methods Enzymol.*, **233**, 363–371.

Griffiths, H.R., Lunec, J., Gee, C.A. and Willson, R.L. (1988a) Oxygen radical induced alterations in polyclonal IgG. *FEBS Letts*, **230**, 155–158.

Griffiths, H.R., Unsworth, J., Blake, D.R. and Lunec, J. (1988b) Oxidation of amino acids within serum proteins. In: Rice Evans, C. and Dormandy, T. (eds) *Free radicals: Chemistry, Pathology and Medicine*, pp. 439–454. Richelieu Press, London.

Griffiths, H.R., Lunec, J. and Blake, D.R. (1992) Oxygen radical induced fluorescence in proteins; identification of the fluorescent tryptophan metabolite, N-formyl kynurenine, as a biological index of radical damage. *Amino Acids*, **3**, 183–194.

Hibbs, J.B., Taintor, R.R., Vavrin, Z. and Rachlin, E.M. (1988) Nitric-oxide – a cytotoxic activated macrophage effector molecule. *Biochem. Biophys. Res. Comm.*, **157**, 87–94.

Johnstone, A. and Thorpe, R. (eds) (1987) Immunochemistry in Practice. Affinity Chromatography and Immunoprecipitation, pp 207–240. Blackwell Scientific Publications.

Jones, A.F. and Lunec, J. (1987) Protein fluorescence and its relationship to free radical activity. *Brit. J. Cancer*, **55**, supplement VIII, 60–65.

Jones, L.A., Holmes, J.C. and Seligman, R.B. (1956) Spectrophotometric studies of some 2,4-dinitrophenyl hydrazones. *Anal. Chem.*, **28**, 191–198.

Kaur, H. and Halliwell, B. (1994) Evidence for nitric oxide-mediated oxidative damage in chronic inflammation – nitrotyrosine in serum and synovial-fluid from rheumatoid patients. *FEBS Letts*, **350**, 9–12.

Kikugawa, K., Kato, T., Beppu, M. and Hayasaka, A. (1991) Development of fluorescence and crosslinks in eye lens crystallin by interaction with lipid peroxyl radicals. *Biochim. Biophys. Acta*, **1096**, 108–114.

Kinter, M., Robinson, C.S., Grimminger, L.C. *et al.* (1994) Whole-blood and plasma-concentrations of 4-hydroxy-2-nonenal in watanabe heritable hyper-lipidemic versus new-zealand white-rabbits. *Biochem. Biophys. Res. Comm.*, **199**, 671.

Levine, R.L., Williams, J.A., Stadtman, E.R. and Shacter, E. (1994) Carbonyl assays for determination of oxidatively modified proteins. *Methods Enzymol.*, **233**, 346–357.

Lunec, J., Blake, D.R., McCleary, S.J. *et al.* (1985) Self-perpetuating mechanisms of IgG aggregation in rheumatoid inflammation. *J. Clin. Invest.*, **76**, 2084–2090.

O'Connell, A.M., Gieseg, S.P. and Stanley, K.K. (1994) Hypochlorite oxidation causes cross-linking of LP(a). *Biochim. Biophys. Acta*, **1225**, 180–186.

Penke, B., Ferenczi, R. and Kovacs, K. (1974) A new hydrolysis method for determining tryptophan in proteins. *Anal. Biochem.*, **60**, 45–50.

Shacter, E., Williams, J.A., Stadtman, E.R. and Levine, R.L. (1996) Determination of carbonyl groups in oxidised proteins. In: Punchard, N.A. and Kelly, F. J. *Free Radicals: a practical approach*, pp 159–170. IRL Press, OUP, Oxford.

Shigenaga, M.K., Lee, H.H., Blount, B.C. *et al.* (1997) Inflammation and nox-induced nitration: assay for 3-nitrotyrosine by HPLC with electrochemical detection, *Proc. Natl Acad. Sci.*, **94**, 3211–3216.

Simat, T., Meyer, K. and Steinhart, H. (1994) Synthesis and analysis of oxidation and carbonyl condensation of tryptophan. *J. Chrom.*, **661**, 93–99.

Steinberg, D., Parasarathy, S., Carew, T.E. *et al.* (1989) Beyond cholesterol – modifications of low-density lipoprotein that increase its atherogenicity. *New Eng. J. Med.*, **320**, 915–924.

Tew, D. and Ortiz de Montellano, O. (1988) The myoglobin protein radical – coupling of tyr-103 to tyr-151 in the hydrogen peroxide-mediated cross-linking of sperm whale myoglobin. *J. Biol. Chem.*, **263**, 17880–17886.

Viner, R.I., Krainer, A.G., Williams, T.D. *et al.* (1997) Identification of oxidation-sensitive peptides within the cytoplasmic domain of the sarcoplasmic reticulum Ca2+-ATPase. *Biochemistry*, **36**, 7706–7716.

Ward, A., McBurney, A. and Lunec, J. (1994) Evidence for the involvement of oxygen-derived free-radicals in ischemia-reperfusion injury. *Free Rad. Res.*, **20**, 21–28.

Young, I.S. and Trimble, E.R. (1991) Measurement of malondialdehyde in plasma by high-performance liquid-chromatography with fluorometric detection. *Ann. Clin. Biochem.*, **28**, 504.

Part II Measurement of 8-Oxo Deoxyguanosine

4 Measurement of 8-Oxo-2′-deoxyguanosine in Cellular DNA by High Performance Liquid Chromatography–Electrochemical Detection

Mark D. Evans

INTRODUCTION

Any tissue DNA can be examined for 8-oxo-2′-deoxyguanosine (8oxodG), as a marker of oxidative damage to DNA, and although there are variations in detail, the assay procedure is well established in the literature (Halliwell and Dizdaroglu, 1992). This compound can exist in either a keto (I in Figure 4.1) or enol (II in Figure 4.1) form, hence its designation as 8oxodG or 8-hydroxy-2′-deoxyguanosine (8OHdG), respectively; the former is more prevalent under physiological conditions.

The level of 8oxodG in DNA represents a balance between exposure of DNA to oxidants and DNA repair processes (Floyd, 1990). DNA extraction

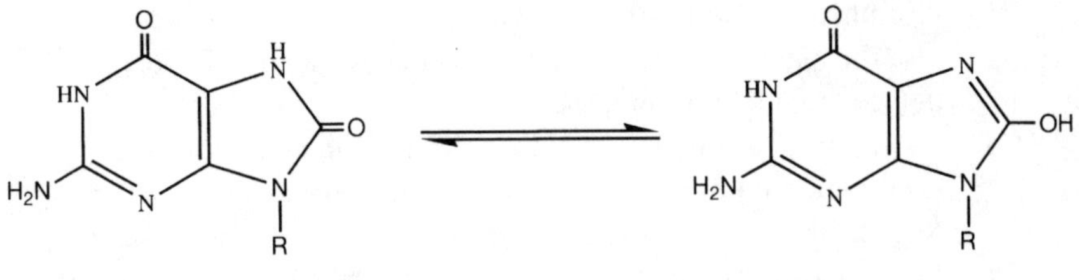

I. 8-OXO-2′-DEOXYGUANOSINE

II. 8-HYDROXY-2′-DEOXYGUANOSINE

Figure 4.1 Keto (**I**) and enol (**II**) forms of 8-oxo-2′-deoxyguanosine; R = 2′-deoxyguanosine moiety.

Measuring in vivo *Oxidative Damage: A Practical Approach.* Edited by J. Lunec and H. R. Griffiths. © 2000 by John Wiley & Sons, Ltd. ISBN 0 471 81848 8.

and enzymic DNA digestion/analysis by HPLC–electrochemical detection (HPLC–EC) are addressed in this chapter.

DNA EXTRACTION

Various DNA extraction procedures have been described in the literature. In many of these procedures whole DNA is extracted, but protocols have been designed to enrich extractions with nuclear DNA at the exclusion of mito-chondrial DNA. The selection of a particular protocol in this regard depends on requirements, bearing in mind some reports suggesting that mitochondrial DNA can contain a higher proportion of oxidized bases than nuclear DNA (Zastawny *et al.*, 1998) . Several of these DNA extraction protocols also rely on the combined use of protease/phenol/organic solvent extractions, but in the interests of limiting artefactual base oxidation and reducing the hazards of the extraction procedures many recent protocols have avoided the use of phenol in particular. The procedures for DNA extraction and the contribution that these extraction procedures might make to the production of artefactual base oxidation are areas of active study (Helbock *et al.*, 1998; Kasai, 1997; Nakae *et al.*, 1995) . As a result these procedures might change in the near future and it is difficult to outline a standard protocol; the reader is advised to refer to the literature regularly to gain information on the latest, critically evaluated DNA extraction protocols. Some general guidelines for selection of the most appropriate DNA extraction methods are presented below.

1. Use ultrapure water throughout.
2. Use the highest purity reagents and enzymes available; free of detectable deoxyribonuclease activity.
3. Use appropriate concentrations of transition metal ion chela-tors, which complex metal ions in a non-redox active form (e.g. deferroximine mesylate) during the extraction procedure.
4. Since the HPLC separation method described (see below) is for the analysis of deoxynucleosides, no ribonuclease digestion steps are required. The inclusion of ribonuclease digestion steps makes the extracted nucleic acid amenable to analysis by a procedure that involves digestion of the DNA to its constituent bases, such as gas chromatography–mass spectrometry.

5. It is recommended that extracted DNA is not freeze-dried or dried by blowing a stream of nitrogen (argon is preferable) over the DNA since this may contribute to artefactual oxidation.
6. Dissolve extracted DNA in the buffer described (see below), rather than water.

Once DNA has been extracted:

1. Dissolve the DNA on a rotary mixer, at 4°C in 10 mM Trizma base (pH 7.5).
2. Analyse the sample for DNA content (A_{260}) and purity (A_{260}/A_{280}) using UV absorbance and store the samples at $-80°C$ if digestion and analysis is to be done later.

QUANTITATION OF DNA

The DNA content and purity of the extracted DNA is analysed using UV absorbance.

1. Dilute the DNA sample to give an absorbance value of between 0.4 and 1.0 unit at 260 nm.
2. Measure the UV absorbance at 260 nm and 280 nm.

Calculations

- DNA purity: Absorbance 260 nm/Absorbance 280 nm = 260/280 ratio
- DNA concentration: μg/ml DNA = A_{260} × dilution factor × 50

1. 50 μg/ml double-stranded DNA has an absorbance of 1 unit at 260 nm.
2. Concentrations of DNA $\geq$ 5 μg/ml can be quantitated by the UV absorbance method.
3. The magnitude of the 260/280 ratio gives an indication of sample purity; a pure preparation of double-stranded DNA gives a value of 1.8.

4. Although this method of DNA quantitation can be inaccurate, particularly if protein contamination is high or the tissue used yields a coloured sample, the values obtained are adequate to gauge the volumes of solution to use for digestion (Laws and Adams, 1996).

DNA DIGESTION AND ANALYSIS OF 8OXODG

DNA is digested to the 2'-deoxynucleoside level using two enzymes, nuclease P_1 and alkaline phosphatase. Both 2'-deoxyguanosine (dG) and 8oxodG are measured in a single run, the former measured by ultraviolet absorbance at 254 nm and the latter by electrochemical detection. Due to the exceptionally low baseline levels of 8oxodG in DNA, electrochemical detection is critical to the analytical procedure.

Reagents

All reagents are prepared in ultrapure water, which is used throughout the following procedures. This water should be depleted of organic/ionic contaminants, and filtered through a 0.2 μm filter to remove microbiological contamination (resistance $\geq 18 M\Omega$).

- Sodium acetate containing zinc chloride: 1.0 M sodium acetate in water containing 45 mM zinc chloride, pH 4.8.
- Desferrioxamine mesylate: 12 mM in water; aliquots can be stored at $-20°$C.
- Trizma base: 1.5 M in water, pH 8.0.
- Nuclease P_1 (EC 3.1.30.1; from *P. citrinum*):

  –Dissolve at ~1100 units/ml in 20–25 mM sodium acetate, pH 4.8.
  –Measure out 7 μl aliquots and store at $-20°$ C.

- Alkaline phosphatase (EC 3.1.3.1; from bovine intestine):
  –Prepare on the day of analysis.

  –Dilute the alkaline phosphatase in 100 mM Tris-HCl, pH 8.0, to give ~750units/ml (note: this is Boehringer–Mannheim units).

- HPLC mobile phase: 50 mM sodium acetate containing 10% v/v methanol, pH 5.1.

1. When selecting a commercial source of nuclease P_1 and alkaline phosphatase, each enzyme solution should be analysed at the appropriate digestion concentration by HPLC–EC. Only select those sources of each enzyme that do not give peaks in either the UV or EC chromatograms that could interfere with dG and 8oxodG derived from DNA. As a guide, in our laboratory we have used nuclease P_1 supplied by Calbiochem-Novabiochem and alkaline phosphatase supplied by Boehringer-Mannheim.

2. For the mobile phase: the use of electrochemical-grade reagents and water is highly recommended; acid and alkali used for pH adjustment should be dedicated for use with mobile phase preparation.

3. Particularly for coulometric detection we recommend these additional guidelines:

   –Solvent bottles and mobile phase filtration apparatus should be washed with electrochemical grade water, methanol and left to dry. These containers should be kept capped/covered when not in use.
   –Mobile phase should be filtered through a $0.2\,\mu m$ filter before use.
   –When adjusting pH, remove an aliquot from the main mobile phase solution and check pH, add alkali or acid as necessary, remove another aliquot and check pH, etc. This avoids introducing the pH probe into the mobile phase.
   –The HPLC pump and injection systems should be scrupulously clean.

HPLC standards

- $2'$-Deoxyguanosine: 1 mM in water.
- 8-oxo-$2'$-deoxyguanosine: 100 μM in water.

Both of these compounds are commercially available.

1. Determine the exact concentration of the standards using UV absorbance and the following extinction coefficients:

–dG, $\varepsilon_{253} = 13.0$ mм/cm

–8-oxodG, $\varepsilon_{245} = 12.3$ mм/cm

3. Aliquot out 0.2 or 0.3 ml volumes of the standards into Eppendorf tubes and freeze-dry.
4. Store the freeze-dried standards at $-20°$C and reconstitute on the day of analysis.

Calibration standards

1. Reconstitute the freeze-dried standards in water using the same volume that was used for freeze-drying.
2. Re-determine the concentration of the standards using UV absorbance as described above.
3. Dilute the 8oxodG solution to 500 nM with water and then measure aliquots of dG and 8oxodG according to the following table to form the standards for the standard curve:

StandardM	1mм dG (μl)	500 nм 8oxodG (μl)	Water (μl)
S0	0	0	200
S1	5	2	193
S2	15	6	179
S3	30	12	158
S4	50	20	130

1. The final volume of these standards will be 238 μl if the digestion procedure is used as stated below. The exact concentrations of the standards can then be calculated based on the UV absorbance measurements for the reconstituted standards.
2. The standard curve noted in the table can be adjusted if it is suspected that the samples to be analysed are substantially damaged, or lower amounts of DNA are used, for example if using coulometric EC detection.

Procedure

1. Aliquot a volume of DNA solution equivalent to 50–70 μg DNA into a suitable capped tube and make up the volume to 200 μl with water.
2. Add 5 μl of 1 M sodium acetate containing 45 mM zinc chloride, pH 4.8, and 2 μl 12 mM desferrioxamine mesylate in water; briefly vortex mix.
3. Add 6 μl of nuclease P_1 to each sample and briefly vortex mix. Incubate the samples at 37° C for 1 h.
4. Add 20 μl of 1.5 M Tris-HCl, pH 8.0; briefly vortex mix.
5. Add 5 μl of alkaline phosphatase to each sample; briefly vortex mix and then incubate at 37°C for 30 min.
6. Either centrifuge the samples at $\sim$ 10 000 rpm in a microfuge (rotor radius of $\sim$ 6 cm) or filter samples through a disposable HPLC sample filter unit.
7. Analyse by HPLC–EC.

1. The enzyme solutions detailed above are designed for use with the amounts of DNA suggested (i.e. 50–70 μg). If it is decided to use a different amount of DNA, the concentrations of nuclease P_1 and alkaline phosphatase stock solutions that are prepared can be modified by taking into account the following minimum requirements:

   – nuclease P_1 at $\sim$ 0.1 units/μg DNA
   – alkaline phosphatase at 0.05–0.075 units/μg DNA.

2. If more than 200 μl of DNA solution is required to give 50–70 μg DNA, then use 200 μl exactly, since this may still yield a measurable 8oxodG peak after digestion.
3. The standards should be processed through exactly the same digestion procedure as the DNA samples, i.e. a total of 200 μl standard containing dG/8oxodG is processed through the same buffer and enzyme additions as the DNA samples.
4. It is important that the digestion mixture be made alkaline *before* adding the alkaline phosphatase.

HPLC–EC

The HPLC procedure described allows separation of all four major 2'-deoxynucleosides in the order 2'-deoxycytidine (dC), dG, 2'-deoxythymidine (dT) and 2'-deoxyadenosine (dA). The 8oxodG peak will elute shortly after the dT peak.

- Column: 150 or 250 mm × 4.6 mm i.d.; stationary phase is 3 μm octadecylsilane (ODS)
- Flow rate: 1 ml/min
- Temperature: ambient
- Injection: 60 μl in duplicate
- Detection parameters:

  – UV detection at 254 nm
  – EC detection at +600 mV vs Ag/AgCl

1. It is recommended that the digests be analysed within 24 hours.
2. Amperometric or coulometric detection modes for the EC can be used. The use of coulometric detection enhances the limit of detection, allowing the use of less DNA per digestion and a lower applied potential. The EC detection parameters reported above are for a standard amperometric detector. The use of a so-called 'coularray detector' has the added advantage of on-line voltammetric information which more robustly identifies the 8oxodG peak.

CALCULATIONS

From the standard curves of peak area vs standard concentration, derive the concentrations of dG and 8oxodG in the samples:

$$\text{Moles on column} = \frac{\text{molarity} \times \text{volume injected (ml)}}{1000}$$

$$\frac{\text{Moles 8oxodG on column} \times 10^5}{\text{Moles dG on column}} = 8\text{oxodG}/10^5\text{dG}$$

The amount of 8oxodG in a sample is usually reported as a ratio relative to dG in order to compensate for differences in digestion efficiency between samples.

ACKNOWLEDGEMENTS

The methodology described in this chapter was designed in part with input from members the European Standards Committee on Oxidative DNA Damage (ESCODD), a group formed to perform inter-laboratory validation of methods for the measurement of oxidative DNA damage. Further contact details relating to this group can be found on-line at http://www.le.ac.uk/pathology/dcp/index.html. The author thanks the following members of the Division of Chemical Pathology: Dr I.D. Podmore for critical reading of the manuscript and Miss N. Mistry for advice concerning coulometric detection guidelines.

REFERENCES

Floyd, R.A. (1990) The role of 8-hydroxyguanine in carcinogenesis. *Carcinogenesis*, **11**, 1447–1450.

Halliwell, B. and Dizdaroglu, M. (1992) The measurement of oxidative damage to DNA by HPLC and GC/MS techniques. *Free Radical Res. Comms.*, **16**, 75–87.

Helbock, J.H., Beckman, K.B., Shigenaga, M.K., Walter, P.B., Woodall, A.A., Yeo, H.C. and Ames, B.N. (1998) DNA oxidation matters: the HPLC–electrochemical detection assay of 8-oxo-deoxyguanosine and 8-oxoguanine. *Proc. Natl Acad. Sci. USA*, **95**, 288-293.

Kasai, H. (1997) Analysis of a form of oxidative DNA damage, 8-hydroxy-2'-deoxyguanosine, as a marker of cellular oxidative stress during carcinogenesis. *Mutation Res.*, **387**, 147–163.

Laws, G.M. and Adams, S.P. (1996) Measurement of 8-OHdG in DNA by HPLC/ECD: the importance of DNA purity. *BioTechniques*, **20**, 36–38.

Nakae, D., Mizumoto, Y., Kobayashi, E., Noguchi, O. and Konishi, Y. (1995) Improved genomic/nuclear DNA extraction for 8-hydroxydeoxyguanosine analysis of small amounts of rat liver tissue. *Cancer Lett.*, **97**, 233–239.

Zastawny, T.H., Dabrowska, M., Jaskolski, T., Klimarczyk, M., Kulinski, L., Koszela, A., Szezesniewicz, M., Stiwinska, M., Witkawski, P. and Olinski, R. (1998) Comparison of oxidative base damage in mitochondrial and nuclear DNA. *Free Radical Biol. Med.*, **24**, 722–725.

5 Immunochemical Detection of 8-Oxodeoxyguanosine in DNA

Marcus Cooke and Karl Herbert

INTRODUCTION

Free radicals and reactive oxygen species are proposed to have an important role in a number of pathological conditions, which include inflammation (Frenkel *et al.*, 1993), autoimmunity (Lunec *et al.*, 1994; Cooke *et al.*, 1997) and carcinogenesis (Floyd *et al.*, 1990). Interaction of these damaging species with cellular macromolecules may lead to modifications with possibly deleterious consequences. DNA is an important target for radical damage, giving rise to many base- and sugar-derived products. Most studied of all the oxidative base lesions is 8-oxo-2′deoxyguanosine (8-oxodG), shown to be potentially mutagenic (Shibutani *et al.*, 1991) and a generally accepted marker of both oxidative stress and oxidative DNA damage. For this reason the identification and quantitation of this lesion in DNA and other biological matrices is essential and a number of assays have been developed to perform this task. Such assays include GC–MS (Dizdaroglu, 1994), HPLC (Floyd *et al.*, 1986; Herbert *et al.*, 1996) and ^{32}P-postlabelling (Podmore *et al.*, 1997); however, these techniques are either technically demanding (either of personnel or equipment) or time-consuming and cumbersome.

Antibody-based technology represents an alternative to these techniques either through immunohistochemical localization of 8-oxodG in the DNA of tissue sections (Hattori, *et al.*, 1997), or as the free deoxynucleoside in cell culture supernatants, serum, plasma or urine (kit). Although monoclonal antibodies to 8-oxodG have been previously applied to the quantitation of 8-oxodG in hydrolysed DNA (Yin *et al.*, 1995), this required the immuno-affinity purification of the hydrolysates prior to competitive ELISA. The resulting assay provided a means of determining the relative amounts of oxidative damage in clinical epidemiology studies. Such studies often involve

Measuring in vivo *Oxidative Damage: A Practical Approach.* Edited by J. Lunec and H. R. Griffiths. © 2000 by John Wiley & Sons, Ltd. ISBN 0 471 81848 8.

investigations of large cohorts and may be longitudinal in nature; therefore there may be many hundreds, or even thousands, of samples for analysis. Simplicity and robustness, in addition to the ability to handle large numbers of samples, makes an immunoassay a possible solution to the problem.

IMMUNOCHEMICAL DETECTION OF 8-OXODG IN DNA

An ELISA for detection of 8-oxodG in urine has been applied in order to detect *in vivo* oxidative stress in human subjects (Erhola *et al.*, 1997). We have adapted a commercially available version of the ELISA to the specific measurement of 8-oxodG derived from enzymically hydrolysed DNA. The immediate advantage over the previously described assay (Yin *et al.*, 1995) is that the assay can be performed without the need for a preliminary immunoaffinity purification of the deoxynucleoside.

Assay principle

The assay may be employed to measure oxidative damage to cellular DNA. Following cell isolation, DNA is purified and subsequently digested to yield deoxynucleosides. The amount of 8-oxodG is then determined by competitive ELISA. Quantitation of the 8-oxodG is dependent upon measuring the coloured end points resulting from assay of the samples and standards of known 8-oxodG concentration; since this is a competitive assay, the greater the amount of 8-oxodG within the test, the lower is the absorbance value.

ELISA dynamic range and limit of detection

3.5 8-oxodG/10^6 DNA bases to 10 845.8 8-oxodG/10^6 DNA bases (based on the DNA concentration in the hydrolysate being 207 μg/ml)

General problems in the measurement of 8-oxodeoxyguanosine

Following collection of samples it is strongly advised that processing be carried out immediately without storage of cells, in order to minimize artefact generation or repair of DNA damage. The former is of particular concern during isolation of DNA from cells. Due to the potential for autooxidation and subsequent generation of 8-oxodG as an oxidation artefact, great care should be taken during DNA isolation to reduce such effects. The procedure described in Chapter 4 is currently under inter-laboratory validation by ESCODD. Once isolated, again it is recommended that the DNA is digested without delay following the procedure described in Chapter 4, using nuclease P_1 (EC 3.1.30.1) and alkaline phosphatase (EC 3.1.3.1) to yield 2'-deoxynucleosides (dG). Measurement of DNA content by measurement of UV absorbance at 260 nm is not accurate given significant variation in protein contamination. The latter is usually assessed by measuring the ratio 260 nm/280 nm. HPLC, GC MS or capillary electrophoresis is required for accurate quantitation of dG present in digests. In our laboratory we routinely use HPLC for this purpose.

ASSAY PROCEDURE

Enzymatic hydrolysis of DNA

Protocol
For full details, see Chapter 4. Briefly, 50 μg samples of DNA in sodium acetate (pH 4.8) are incubated at 37°C with nuclease P1 for one hour. After adjusting the pH to 8.0, the samples are then incubated with alkaline phosphatase at 37°C for 30 min, prior to HPLC analysis. Deoxyguanosine and 8-oxodeoxyguanosine standards are treated in an identical manner to the DNA samples.

Quantitation of deoxyguanosine in DNA digests

Protocol
For full details, see Chapter 4. Briefly, 60 μl volumes of hydrolysed samples and deoxyguanosine standards (0–200 μM) are injected on to a Hypersil ODS (C_{18}) column (150 mm $\times$ 4.6 mm) and the deoxyguanosine is detected at 254 nm. Injection and detection are performed by a Beckman HPLC system.

The peak area data for the standards is used to produce a plot of peak area versus concentration, from which dG values of the samples are determined.

ELISA analysis of 8-oxodG in DNA

This analysis takes 3.5 to 4 h to complete.

Reagents
- Japan Institute for the Control of Aging 8-oxodG ELISA kit (Genox Corporation, 1414 Key Highway, Baltimore, MD 21230, USA) which contains: monoclonal antibody to 8-oxodG, peroxidase-labelled anti-mouse secondary antibody, 96-well ELISA plate coated with 8-oxodG and standard solutions of 8-oxodG (0.64, 3.2, 16, 80, 400 and 2000 ng/ml).
- Tween 20 (0.05% v/v).
- Phosphate-citrate buffer (0.05 M), containing 0.03% perborate buffer.
- Orthophenylenediamine, 0.5 mg/ml (care: light sensitive).
- Sulphuric acid (2 M).

Procedure
1. All reagents and ELISA plates must be allowed to warm to room temperature (20–25°C) before use.
2. Reconstitute primary antibody with primary antibody solution as supplied with the kit.
3. Add standard solutions of 8-oxodG (0.64–2000 ng/ml) and sample DNA hydrolysates to the plate in triplicate, 50 μl/well. Blank wells, consisting of sample, secondary antibody and chromatic substrate, i.e. no primary antibody.
4. Add 50 μl of the primary antibody (0.2 μg/ml final concentration) to all wells and incubate plate sealed with adhesive strip at 37°C for 1 h.
5. Following incubation, empty the contents of the plate and thoroughly wash plate with 200 μl/well of wash buffer (0.05% v/v Tween 20 in 0.01 M PBS, pH 7.4). Invert the plate and blot.
6. Add the secondary antibody, 100 μl/well, and incubate for 1 h at 37°C.
7. Wash plate as described in step 5.
8. Prepare the substrate solution as follows: orthopenylenediamine (0.5 mg/ ml in 0.05 M phosphate-citrate, pH 5.0, and containing 0.03% sodium perborate). This is added to the 96-well plate (100 μl/well) with incubation for 15 min, in the dark, at room temperature.
9. Stop the reaction sulphuric acid (2N) using 25 μl/well before reading at 492 nm.

Analysis of results

Plotting absorbance versus $\log_{10}$ concentration of the standards produces a calibration curve from which can be determined the amount of 8-oxodG in the test samples. The amount of DNA added to the plate in the hydrolysates is calculated from the concentration of deoxyguanosine determined by HPLC or capillary electrophoresis.

Calculation

From the 8-oxodG calibration curve can be calculated the concentration of 8-oxodG (ng/ml):

- x ng/ml $\times$ 0.06 = amount of 8-oxodG per volume of digest injected on to the HPLC, i.e. 60 μl (in nanograms)
- Multiplication of this by 3.53×10^{-3} converts this value to number of nanomoles.

HPLC analysis of the DNA digests provides levels of deoxyguanosine in nanomoles per volume injected (60 μl). The ratio between 8-oxodG (nmol) and dG (nmol) will give a value of 8-oxodG/10^5 dG.

Comparison with an established HPLC technique

Baseline values for calf thymus DNA were determined using this ELISA method (3.7–6.8 8-oxodG/10^5 dG) and compared with levels measured by HPLC–ECD in this laboratory (3.06–3.78 8-oxodG/10^5 dG).

Limitations of the assay

If the 8-oxodG content of the samples analysed exceeds the range of the standard curve, the samples must be diluted and the assay repeated. Conversely, samples with a low 8-oxodG concentration may require concentrating.

REFERENCES

Cooke, M.S., Mistry, N., Wood, C.W. *et al.* (1997) Immunogenicity of DNA damaged by reactive oxygen species – implications for anti-DNA antibodies in Lupus. *Free Radical Biol. Med.*, **22**, 151–159.

Dizdaroglu, M. (1994) Chemical determination of oxidative DNA damage by GC–MS. *Methods Enzymol.*, **234**, 3–16.

Erhola, M., Toyokuni, S., Okada, K. *et al.* (1997) Biomarker evidence of DNA oxidation in lung cancer patients: association of urinary 8-hydroxy-2'-deoxyguansoine excretion with radiotherapy, chemotherapy and response to treatment. *FEBS Letts*, **409**, 287–291.

Floyd, R.A., Watson, J.J., Wong, P.K *et al.* (1986) Hydroxyl free radical adduct of deoxyguanosine: sensitive detection and mechanism of formation. *Free Radical Res. Comm.*, **1**, 163–172.

Floyd, R.A. (1990) The role of 8-hydroxyguanosine is carcinogenesis. *Carcinogenesis*, **11**, 1447–1450.

Frenkel, K., Karkoszka, J., Kim, E. *et al.* (1993) Recognition of oxidised DNA bases by sera of patients with inflammatory diseases. *Free Radical Biol. Med.*, **14**, 483–494.

Hattori, Y., Nishigori, C., Tanaka, T. *et al.* (1997) 8-hydroxy-2'-deoxyguanosine is increased in epidermal cells of hairless mice after chronic ultraviolet B exposure. *J. Invest. Dermatol.*, **107**, 733–737.

Herbert, K.E., Evans, M.D., Finnegan, M.T.V. *et al.* (1996) A novel HPLC procedure for the analysis of 8-oxoguanine in DNA. *Free Radical Biol. Med.*, **20**, 467–473.

Lunec, J., Herbert, K., Blount, S. *et al.* (1994) 8-Hydroxy deoxyguanosine: a marker of oxidative DNA damage in systemic lupus erythematosus. *FEBS*, **348**, 131–138.

Podmore, K., Farmer, P.B., Herbert, K.E. *et al.* (1997) ^{32}P-Postlabelling approaches for the detection of 8-oxo-2'-deoxyguanosine-3'-monophosphate in DNA. *Mutation Res.*, **178**, 139–149.

Shibutani, S., Takeshita, M. and Grollman, A.P. (1991) Insertion of specific bases during DNA synthesis past the oxidation-damaged base 8-oxodG. *Nature*, **349**, 431–434.

Yin, B., Whyatt, R.M, Perera, P. *et al.* (1995) determination of 8-hydroxydeoxyguansoine by an immunoaffinity chromatography–monoclonal antibody-based ELISA. *Free Radical Biol. Med.*, **18**, 1023–1032.

6 Urinary measurement of 8-OxodG (8-Oxo-2'-deoxyguanosine)

Henrik E. Poulsen, Steffen Loft and Allan Weimann

INTRODUCTION

The most abundant oxidative modification known in DNA is 8-hydroxylation of the guanine moiety. This modification is also mutagenic, preferentially resulting in G-T transversion mutation, demonstrated, for example, in the codon 248 of tumour supressor gene *P53*. Due to these observations, urinary excretion of the repair product 8-oxodG has attracted interest as a potential biomarker for the development of certain cancers (Loft and Poulsen, 1996).

Potential problems

- It has yet to be verified that urinary excretion of 8-oxodG is prognostic for development of cancer.
- Defined period of urine collection (24 hour or similar) is necessary for estimation of the excretion rate.
- Correction of spot urine samples by creatinine concentration does not correlate well with 24-hour urine excretion. Under certain circumstances (short trials, repeated measurements in the same individual, unchanged diet and muscle mass) short urine collection periods or even spot urine may be applicable. It is evident that group comparison (e.g. healthy versus diseased persons) cannot be done by spot urine and creatinine

Measuring in vivo *Oxidative Damage: A Practical Approach.* Edited by J. Lunec and H. R. Griffiths. © 2000 by John Wiley & Sons, Ltd. ISBN 0 471 81848 8.

comparison, because of the large difference in creatinine excretion in the groups.

- Blank urine, i.e. urine without 8-oxodG, is not available and estimation of the urinary concentration must rely on sample addition or labelled internal standard.

- At present, ELISA methods do not appear to correlate with other methods. Values are high presumably due to insufficient specificity of the antibodies available.

- Synonyms for 8oxodG are: 7,8-dihydro-8-oxo-2'- deoxyguanosine, 8-oxo-2'-deoxyguanosine, 8-hydroxydeoxyguanosine, 8-hydroxy-2'-deoxy- guanosine, 8OHdG, oxo^{8}dG.

- It is not known if individual risks can be assessed by measurement of urinary 8-oxodG, or whether the potential as biomarker is limited to group comparison.

- The methodology for estimation of urinary 8xodG can also be used to estimate levels in tissue DNA. Particular care should be taken to avoid artificial oxidation in this case. For urine this problem is not of the same magnitude, due to low levels of non-oxidized bases and nucleotides.

METHODS OF ANALYSIS

The published values of urinary excretion of 8oxodG range from 110 to 600 pmol/kg body weight/24 h, and urinary concentrations range from about 1 nM to about 100 nM. There is no accreditation procedure or quality assurance established and the method is purely for research purpose.
The published methodologies include:

- enzyme-linked immuno-sorbent assay (ELISA)
- gas chromatography with selective ion monitoring (GCMS–SIM)
- high performance liquid chromatography with electrochemical detection (HPLC–EC)
- liquid chromatography with tandem mass spectroscopy (LCMS–MS)

The choice between the different methodologies depends mainly on the apparatus and know-how in the laboratory to set up the analysis, and the funds available.

The ELISA method is attractive because of the low price and simple technological requirements and because it relies on any brand of ELISA plate reader. The assays available so far overestimate levels by 3–5 times and at present the method cannot be recommended (Prieme *et al.* 1996). It is not further described in this chapter.

GCMS has been used in very few laboratories and is mainly used for estimation of 8-oxoGua in tissue DNA (see Chapter 4). HPLC–EC is the preferred methodology and is therefore described in detail.

LCMS–MS is a promising methodology, with the potential advantages of rapid analysis, simultaneous determination of other oxidative DNA base/nucleotide modifications and a sensitivity matching that of HPLC–EC. It has the advantage over GCMS of eliminating the derivatization step, thereby reducing the potential for introducing artificial oxidation.

HPLC–EC ANALYSIS

Apparatus and system set-up

- Two high-pressure HPLC pumps (Merck-Hitachi L-6000 and L- 6200)
- Column oven (40°C)
- Pulse dampener (LP21, Science Systems Inc.)
- Six-port fast switching valve (Valco)
- Electrochemical detector (ESA Coulochem II with 5011 high sensitivity cell)
- UV detector (Waters 440 UV detector, 254 nm)
- Autoinjector (Merck-Hitachi 655A-40)
- Extraction column (Spherisorb ODS2, 15 cm, 5 μm, Waters Denmark)
- Cation exchange (CE) column, 2 cm (Hamilton, Reno, Nevada)
- Analytical column (Nucleosil C18, 3 μm, Knaur, Germany)
- PC chromatographic data handling and integration software system (Merck-Hitachi D-6000 or higher).

System set-up is indicated in Figure 6.1. The six-port switching valve should be set to give two situations:

1. Flow from pump 1 is: extraction column → valve → CE column → UV detector → waste. Flow from pump 2 is: valve → analytical column → EC detector → waste.
2. Flow from pump 1 is: extraction column → valve → UV detector → waste. Flow from pump 2 is: valve → CE column → analytical column → EC detector.

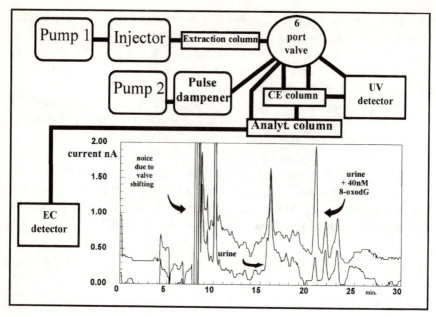

Figure 6.1   HPLC–EC analysis system set-up

- The brand of the HPLC hardware components can be chosen ad lib. However, in our experience the electrochemical detector brand suggested is the most stable and reliable among the ones tested and it gave the most reproducible results.
- When setting up a new cell for electrochemical detection, record a dynamic voltamogram (160–400 mV) by injecting an 8-oxodG sample repeatedly and changing the voltage on electrode 2 in increasing steps for recording the response on the detector. The voltamogram is an S-curve giving the relationship between applied voltage and response; setting of working potential should correspond to the first part of the upper part of the S-curve. This procedure should be repeated at regular intervals. Deterioration of the cell is seen as a right displacement of the voltamogram.
- The 5011 cell has two voltages, to be set corresponding to a first and second set of electrodes. Set the first voltage at about

100–120 mV and record the signal from the second electrode according to the voltamogram (this will reduce noise from substances that could give background interference from being electrochemically active at low voltage).

- Flushing of the cell with 200 μM ascorbic acid can revive the cell to some extent, but replacement is needed when the voltamogram does not show a plateau or shifts to the right.

The HPLC system is started with the six-port switching valve set for situation (1). Urine spiked with 2–4000 nM 8oxodG solution is injected for determination of the retention time until elution from the extraction and CE columns on to the UV detector. The high concentration is used to give a detectable signal on the UV detector. An electrochemical detector can be used instead of the UV detector but is not as cost-effective.

When the retention time of 8oxodG from the CE column is determined, the system is set to run in situation (1) until 1 min prior to elution; it is then switched to situation (2) for 1 min, after which it is reset to situation (1) for the remainder of the run. This will ensure that only the peak of interest (i.e. the 8oxodG peak) is diverted to the analytical column. The shift in flow/pressure during the valve shifting gives rise to large signals on the EC detector; however, this does not disturb the quantification of the 8-oxodG peak. The urine samples may give some variation in the retention time of the CE column and must be checked and set individually.

Quantification is by spiking the urine sample to be analysed with two different concentrations of pure 8oxodG.

Calculation
The 8-oxodG peak height P1 are recorded after injection of pure urine; P2 and P3 are recorded after urine has been spiked with 8-oxodG. .

The concentration in the pure urine sample is calculated twice, from each of the spiked samples:

$$C1 = P1 \times (P2 - P1)/20$$
$$C2 = P1 \times (P3 - T1)/200$$

The average, $(C1 + C2)/2$, is used as the measured concentration; however, C1 and C2 should not differ by more than 10%. A greater difference indicates that separation on the analytical column is not satisfactory and that the mobile phase needs adjustments. Only trial and error can indicate how to adjust the acetonitrile concentration (most often somewhere between 2.5 and 4%). Minor adjustment of pH can also bring about satisfactory separation.

Reagents
- 8oxodG standard (Sigma SKJ 7700, molecular weight 283.2 g/mol)
- Tris (Sigma T-1503 Lot 55H55703)
- Boric acid (Riedel-de Haën, Holland 31146)
- Ortho-phosphoric acid (Riedel-de Haën, Holland 30417)
- Millipore filters (0.45 μm, cat HVLPO 4700 lot R8AM444479)

Mobile phase on pump 1
One litre of mobile phase is made from:

- 25 ml acetonitrile, HPLC grade
- 15 ml methanol, HPLC grade
- 960 ml double distilled water (or HPLC grade)
- 1.237 g sodium borate

Adjust pH with NaOH to 7.9, de-gas and filter through millipore filters (0.45 μm).

Mobile phase on pump 2
One litre of mobile phase is made from:

- 25–40 ml acetonitrile, HPLC grade (adjust as indicated from C1 and C2 concentrations as described above)
- 6.7 ml orthophosphoric acid
- Water, double distilled or HPLC grade: add 1 l (last 10–20 ml added after pH adjustment)

Adjust pH with 5 N NaOH to 2.1 (usually requires about 5 ml for a start).

Preparation of urine samples

Fresh urine, or urine samples stored at −20°C, can be used. No preservation is necessary for storage and 8oxodG is stable for years. The following procedure can be applied before or after storing the sample.

Reagents
- Tris solution, 1 M: 6.055 g in 100 ml double distilled water

Procedure
1. Use 2 ml of urine (fresh or stored).
2. Add 40 μl 2 M HCl to ensure acidic urine, and freeze.
3. Thaw sample and centrifuge at 3000 g for 10 min.
4. Mix 1.7 ml of supernatant with 34 μl 2 M NaOH.
5. Mix 95 μl supernatant and 5 μl 8-oxodG (Sigma, St Louis, Missouri) 0 nM in double distilled water (sample **S1**).
6. Mix 95 μL supernatant and 5 μl 8-oxodG 200 nM in double distilled water (sample **S2**).
7. Mix 95 μl supernatant and 5 μl 8-oxodG 2000 nM in double distilled water (sample **S3**).
8. Mix each of the three samples (**S1, S2, S3**) with 100 μl Tris buffer, pH 7.9, resulting in final spiking with 20 and 200 nM 8-oxodG.
9. Inject 25 μl of each of the three samples on to the HPLC system.

Alternative HPLC–EC procedures

Some researchers have suggested the use of two separate HPLC systems: the first for collection of the peak of interest, which is evaporated and then injected on to a second HPLC system with different elution composition. Variations in the switching system and set-up are also used (Tagesson *et al.*, 1995) and appear to be equally functional as the system described here.

Immunoaffinity columns have also been used (Shigenaga *et al.*, 1994) but such columns are not commercially available and require labelled standards to estimate recovery.

Quality control

The described assay should give linear response curves to injection of standard solutions of 8-oxodG. The inter-assay coefficient of variation should be less than 9–13% and the limit of detection should be about 0.2 nM based on a signal to noise ratio of 1:3.

Complete separation of the 8-oxodG peak can be ensured from the lack of difference between the two calculated concentrations C1 and C2. The procedure is given above under calculations.

Final calculation

Excretion of 24-hour urinary 8-oxodG is calculated by multiplying the estimated urinary concentration of 8-oxodG with the 24–hour urine volume, giving a unit of nM/24 h. The excretion rate can also be measured per kilo-gram body weight, giving a unit of nM/24 h/kg BW.

Urine collection period can be varied; however, it has to borne in mind that the half-life of injected 8-oxodG from plasma is about 2–3 h. Because of this, reduction of collection time to less than 9–12 h cannot be recommended. Other means of 'biological average' can be used; for example, three overnight urine collections, or 48 h collection.

Correction of spot urine samples by estimation of urinary creatinine con-centration cannot be recommended.

LCMS–MS ANALYSIS OF 8-OXODG IN URINE

Introduction

Electrochemical detection is renowned for very high sensitivity compared with UV detection and also sensitivity better or equal to that achievable by mass spectroscopy. HPLC–EC is certainly the most widely applied method for measuring oxidative damage to DNA, but, the methodology also has some limitations. The HPLC–EC method described above requires re-run of the sample and runs of additional spiked samples. Separation of 8-oxodG from other compounds in urine is quite difficult and the complicated set-up with column switching, although automated, requires human surveillance and multiple repeated adjustments. At best the capacity is up to five samples per day, and in our experience it is fewer than 500 samples per working year, even with the dedication of a full HPLC system and a technician. Further-more, the assay is not easily extended to measure other oxidized nucleosides or the oxidized bases.

Gas chromatography–mass spectrometry (GCMS) has been used for mea-suring oxidative DNA modification, but extensive work-up and derivatiza-tion of samples is necessary for this type of analysis. It has been argued and also demonstrated that certain of these procedural steps can induce oxida-tion of non-oxidized DNA bases/nucleosides, giving rise to false high values–particularly in DNA extracts.

Coupling liquid chromatography (LC) with mass spectroscopy is a way to avoid these problems. However, the peaks resulting from LC do not match the height and narrowness of those from GC, and the single quadruple is not sensitive and selective enough to match the sensitivity of HPLC–EC.

To overcome these limitations we have started to develop an LCMS–MS method for analysis of urinary 8-oxodG, with a potential for simultaneous measurement of other excreted DNA repair products (e.g. 8-oxoGua, 8-oxoAdenine and 8-oxodA) and with a potential for measuring DNA samples from extracted tissue.

Two different types of tandem mass spectrometers are available as bench-top machines: triple quadruples and ion trap, the latter with the potential for MS^N mass spectroscopy. Each type provides advantages and disadvantages. Ion trap mass spectrometers are particularly renowned for high sensitivity in full scan mode and are therefore particularly suited for identification purpose. In practice it is difficult to exploit the theoretical MS^N capability for N larger than 2. When it comes to quantification and analysis in SRM (selected reaction monitoring) mode, the triple quadruples are better suited and are renowned for high sensitivity in SRM mode.

For development of the LC part of the urinary assay for 8-oxodG, several columns were tested for qualities such as separation of oxidized and non-oxidized base and nucleoside and relatively short retention time, with a mobile phase without involatile buffers. The best suitable column in our experience is as indicated below.

Instrumentation set-up

Liquid chromatography
• Hewlett-Packard LC model 1050 series (HP1050)

Tandem mass spectrometer
• Perkin-Elmer SCIEX API 365 LC/MS/MS (API365) with Turbo, Ion-spray®, and Perkin-Elmer software running on an Apple Power Macin-tosh 7300/200.

HPLC packing pump
• Magnus Scientific P6060 HPLC Slurry Packer, connected to a 200 ml stainless steel chamber for vertical packing.

A 50 × 2.0 mm i.d. steel column was packed with Waters Spherisorb S3 C6 material from a Waters Spherisorb 4.6 mm i.d. 15 cm column (part no. PSS839999, Batch 1003). Packing can be done either with methanol or (probably better) with acetone.

Preparation of 1 l of mobile phase for the LC
• 20 ml HPLC grade acetonitrile
• 1 ml acetic acid

- Add double distilled water (1 l) and adjust pH to 4.0 with ammonia solution (148 ml ammonia to 1000 ml double distilled water).

Interface between HP1050 and API365

The HP1050 is connected to the S3C6 column eluted at a flow rate of 0.2 ml/min. Methanol (flow 0.1 ml/min) is added to the effluent (flow 0.2 ml/min) before it enters the mass spectrometer (by means of a Merck Hitatchi L-6000 pump). Addition of methanol improves the sensitivity, presumably by increasing ionization.

Experience

The LCMS–MS method is not fully developed at present but sufficiently documented for the suggested set-up to be operational. The scan in Figure 6.2 of a 1000 nM sample (six scans) shows that the dominant product ion from the precursor ion at m/z 284 is m/z 168 from the precursor at m/z 284. This corresponds to selection of the protonated 8-oxodG nucleoside in Q1, breaking of the base–sugar bond NQ2 and selection of the resulting protonated base in Q3.

Figure 6.3 gives mass chromatographs of a normal human urine sample (same urine injected twice, no preparation) and the run of a 1000 nM aqueous

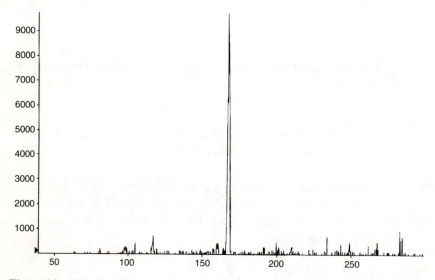

Figure 6.2   LCMS–MS results

8-oxodG sample (dashed line). As indicated by the retention time and the peak shape, there is identity of the genuine sample and the urine measurement. From time 0 to 2.2 min during which 8oxoGua is detected, m/z 168 (the protonated base) is selected in Q1; and m/z 140 (the most intense product ion, scan not given) is selected in Q3 for the remaining time that 8-oxodG is detected.

At present, internal standardization is not implemented and isotope-labelled 8-oxodG is not commercially available. The urine sample in Figure 6.3 corresponds to a concentration of about 40 nM, as judged from injection of aqueous 8-oxodG in concentrations of 1–1000 nM (correlation coefficient of this standard curve is 0.9999). Preliminary investigations on spiked urine samples show a sensitivity loss in urine of about 2.5, which clearly demonstrates the need for internal standardization.

Interpretation of urinary measurements

Estimation of urinary 8-oxodG excretion provides an estimate of whole-body oxidative stress to DNA, and as such is a rate measurement. It is

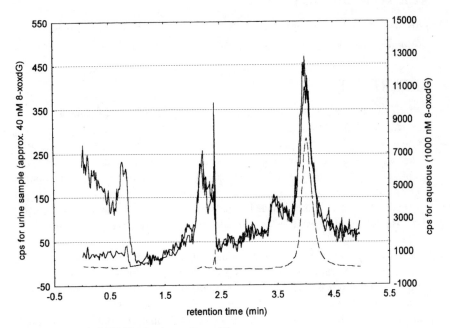

Figure 6.3   LC–MSMS of urinary 8oxodG

independent of DNA repair. This should not be confused with the concentration measurement of DNA oxidation in tissue or cell samples. Choice of measurements must rely on the specific experiment and the hypothesis under investigation. More detailed argumentation is given in Poulsen *et al.* (1998).

Conclusion

LCMS–MS analysis of urine for 8-oxodG is clearly feasible and offers several advantages over the HPLC–EC method: shorter analysis time, high specificity, quantification by internal standardization, and no need for column switching. As such, it offers a high capacity analysis for use in molecular epidemiology. The disadvantage compared with the HPLC–EC method is the four-fold higher price and the need for experience in both LC and tandem mass spectroscopy.

Although not yet demonstrated the method has the potential for measuring other oxidised nucleosides and bases, and also for use in the analysis of DNA samples extracted from tissue.

ACKNOWLEDGEMENTS

This work has been supported by BAT and the Damish Medical Research Council.

REFERENCES

Loft, S. and Poulsen, H.E. (1996) Cancer risk and oxidative DNA damage in man. *J. Mol. Med.,* **74**, 297–312.

Poulsen, H.E., Prieme, H. and Loft, S. (1998) Role of oxidative DNA damage in cancer initiation and promotion. *Eur. J. Cancer Prevent.,* **7**, 9–16.

Prieme, H., Loft, S., Cutler, R.G. and Poulsen, H.E. (1996) Measurement of oxidative DNA injury in humans: evaluation of a commercially available ELISA assay. In: Kumpulainen, J.T. and Salonen, J.T. (eds) *Natural Antioxidants and Food Quality in Atherosclerosis and Cancer Prevention, pp. 78–82. The Royal Society of Chemistry, London.*

Shigenaga, M.K., Aboujaoude, E.N., Chen, Q. and Ames, B.N. (1994) Assays of oxidative DNA damage biomarkers 8-oxo- 2′-deoxyguanosine 8-oxoguanine in nuclear DNA and biological fluids by high performance liquid chromatography with electrochemical detection. *Methods Enzymol.,* **234**, 16–33.

Tagesson, C., Kallberg, M. and Wingren, G. (1996) Urinary malondialdehyde and 8-hydroxydeoxyguanosine as potential markers of oxidative stress in industrial art glass workers. *Int. Arch. Occup. Environ. Health,* **69**, 5–13.

Part III Cellular-based Methods

7 Measurement of Oxidative DNA Damage Using the Comet Assay

Andrew R. Collins

INTRODUCTION

The comet assay is a relatively simple, rapid and sensitive method for measuring DNA strand breaks at the level of single cells. It is increasingly used in genotoxicity testing as well as in human biomonitoring studies. Essentially, a cell suspension (e.g. lymphocytes isolated from venous blood) is mixed with agarose, allowed to set in a thin layer on a microscope slide, and lysed with detergent and high salt to remove cytoplasm, membranes and most nuclear proteins. What remains is 'nucleoids' – supercoiled DNA, arranged in loops attached to the nuclear matrix, morphologically similar to the original nuclei of the cells. The gels are then subjected to alkaline electrophoresis. DNA is attracted to the anode, but only those loops containing a break, which relaxes supercoiling, are free to migrate; they extend from the nucleoid 'head' to form the 'tail' of a comet-like image, viewed by fluorescence microscopy after staining with a suitable dye.

The comet assay has been modified for the specific detection of oxidative DNA damage. The bacterial repair enzyme, endonuclease III, recognizes oxidized pyrimidines, cuts them out, and nicks the DNA at the resulting apurinic/apyrimidinic (AP) site. If nucleoids are incubated with this enzyme, after lysis, the number of relaxed loops, and hence the relative intensity of the tail, is increased according to the number of oxidized pyrimidines present (Collins *et al.*, 1993). Another enzyme, formamidopyrimidine glycosylase, acts similarly but at the site of altered purines, including 8-oxo-guanine,

Note: Mention of a particular manufacturer's product does not imply endorsement of this or other products of that manufacturer.

Measuring in vivo *Oxidative Damage: A Practical Approach.* Edited by J. Lunec and H. R. Griffiths. © 2000 by John Wiley & Sons, Ltd. ISBN 0 471 81848 8.

probably the most abundant base-oxidation product to be found in DNA (Dušinská and Collins, 1996).

We shall describe the comet assay as we use it – a simplified version of the original (Singh *et al.*, 1988) with the addition of the enzyme digestion step. We shall also give an account of methods of quantitation and, in brief, some applications.

REAGENTS AND MATERIALS

Prepare solutions from appropriate stocks, such as 0.5 M Na$_2$EDTA, 1 M Tris, 1 M KCl, etc. Keep solutions at 4°C.

- Microscope slides: use either normal, precleaned glass slides (in which case they must be precoated with agarose and dried; see procedure below), or all-over frosted microscope slides supplied by: Richardson Supply Co., London; Cat. No. 267–096
- Agarose:

  (*a*) Electrophoresis grade (e.g. Gibco BRL 5510UA)
  (*b*) LMP (low melting point) (e.g. Gibco BRL 5517US)

Agarose is dissolved in water or PBS (see below) by heating in an *unsealed* vessel in a microwave oven. Prepare about 10 ml at a time; it can be kept at 4°C for several days. For use, reheat to melt; (*b*) must then be held at 37°C, but (*a*) is liquid only at higher temperature.

- Lysis solution: 2.5 M NaCl, 0.1 M EDTA, 10 mM Tris. Prepare 1 litre. Set pH to 10 with either solid NaOH, or preferably concentrated (10 M) NaOH solution. (Add 35 ml of 10 M NaOH straight away to ensure that EDTA dissolves, and then add dropwise to pH 10.) Keep at 4°C. Add 1 ml Triton X-100 per 100 ml *immediately before use.*
- Enzyme reaction buffer for endonuclease III and FPG: 40 mM HEPES, 0.1 M KCl, 0.5 mM EDTA, 0.2 mg/bovine serum albumin ml; pH adjusted to 8.0 with KOH (this buffer can be made as a 10 × stock, adjusted to pH 8.0 and frozen at −20°C).
- Electrophoresis solution: 0.3 M NaOH, 1 mM EDTA.
- Neutralizing buffer: 0.4 M Tris; pH adjusted to 7.5 with conc. HCl.
- Stain: 4′, 6-diamidine-2-phenylindol dihydrochloride (DAPI), 1 mg/ml in water, further diluted to 1 μg/ml and stored frozen as 1 ml aliquots.
- Also needed: PBS; Histopaque 1077 (lymphocyte separation medium, e.g. from Sigma); cell culture medium, e.g. RPMI (for lymphocytes); MEM (for cultured cells), with/without foetal calf serum (FCS) supplement.

EQUIPMENT

- Horizontal electrophoresis tank: some companies sell tanks specially designed for use in the comet assay, but standard tanks are perfectly satisfactory. A platform 21 cm wide × 20 cm long will accommodate 16 slides arranged in two rows.
- Power unit: ∼ 200 V, 1000 mA.
- Fluorescence microscope: 60 × objective; filters appropriate to the fluorescent DNA stain selected.
- Image analysis system: not essential, but may be reassuring. Supplied by Kinetic Imaging, Liverpool; or Perceptive Instruments, Haverhill, Suffolk; or Laboratory Imaging, Prague.

PROCEDURE

1. Slide preparation

Fully frosted slides give good anchorage for agarose, but are expensive if used only once; unfortunately after a few uses agarose is no longer reliably retained.

Ordinary clear slides (with a frosted end for labelling) must be precoated with agarose by dipping in a (vertical) staining jar of melted 1% standard agarose in H_2O, draining off excess agarose, wiping the back clean and drying in a warm oven.

Advantages: economy; can be stored dry after electrophoresis for later examination (frosted slides give very poor images after drying, as the comets are then so close to the glass that the frosting causes very high background fluorescence).

2. First agarose layer

Optional in the case of precoated slides, *normal* for frosted slides.

Place on the slide 85–100 μl of 1% standard agarose in PBS and, while still liquid, cover with a cover slip (18 × 18 or 22 × 22 mm). Place slides in refrigerator for at least 5 min for agarose to solidify. Push coverslip off first agarose layer with a thumb just before adding top layer (step 4 below). (*Note:* Two gels can be set on each slide if desired; this doubles the number of samples that can be processed in one electrophoresis run.)

3. Preparation of cells

Lymphocytes
Take about 30 μl blood from finger prick, or venous blood sample (+ anti-coagulant). Add to 1 ml PBS (or 1 ml RPMI medium + 10% FCS) in a 1.5 ml Eppendorf tube. Mix and leave on ice for 30 min. Then underlay with 100 μl Histopaque 1077 (Sigma) or similar, using a micropipette. Centrifuge at 200 ×g, 3 min, 4°C. Retrieve lymphocytes in 100 μl from just above boundary between PBS (RPMI) and Histopaque, using micropipette. Add to 1 ml PBS. Spin again. Remove as much supernatant as possible using micropipette.

Cultured (monolayer) cells
Wash cells growing in dish with PBS, and add trypsin/EDTA (usual reagent for harvesting cells); incubate until cells are rounding up, remove trypsin, add 1 ml of appropriate medium, detach cells by pipetting. Alternatively, use a silicon rubber scraper to remove cells. Transfer cells to Eppendorf tube, spin at 200 × g, 3 min, 4°C. Remove supernatant, disperse pellet in 1 ml PBS. Spin again, and remove as much supernatant as possible using micropipette.

Other cell types
Tissues must be disaggregated physically and/or enzymically before the cells can be processed with the comet assay. Generally, cells survive the disaggregation remarkably well, with little sign of non-specific breakage of DNA having occurred during preparation. An exception to the principle of 'single cell electrophoresis' is the analysis of DNA damage in whole crypts of colon epithelium (Brooks and Winton, 1996); multiple comet tails form a 'shadow' on the anode-facing side of the crypt.

4. Embedding cells in agarose

Tap tube to disperse cells in the small volume of medium remaining. Quickly add 140 μl of 1% LMP agarose in PBS at 37°C and mix by tapping tube and then aspirating agarose up and down, using micropipette. Take 140 μl of mixture (use same pipette tip) and transfer as two roughly equal drops on a slide. Cover each with an 18 × 18 mm coverslip. Work quickly, as the agarose sets quickly at room temperature (practice is needed; it is safest to start with 70 μl of suspension to make one gel per slide). Leave slides in refrigerator for 5 min.

5. Lysis

Remember to add 1 ml Triton X-100 to 100 ml lysis solution just before use. Remove coverslips from slides and place in this solution in a (horizontal) staining jar. Leave at 4°C for 1 h.

6. Enzyme treatment (endonuclease III, formamidopyrimidine glycosylase)

Prepare 300 ml of enzyme reaction buffer. Put aside 1 ml for enzyme dilutions. Wash slides in three changes of this buffer (4°C) in staining jar, for 5 min each. Meanwhile, prepare dilutions of enzyme (see Note (B) below).

(*Note:* The buffer in which the enzyme is prepared contains β-mercaptoethanol to preserve the enzyme. However, inclusion of sulphydryl reagents in the reaction buffer would significantly increase background DNA breakage.)

Remove slides from last wash, and dab off excess liquid with tissue. Place 50 μl of enzyme solution (or buffer alone, as control) on to gel, and cover with 22×22 mm coverslip. Put slides into moist box (prevents desiccation) and incubate at 37°C for 45 min (endo III) or 30 min (FPG).

7. Alkaline treatment

Electrophoresis solution should be cooled before use, e.g. by pouring into the electrophoresis tank in the cold room an hour or so before it is needed. Gently place slides (minus coverslips) on platform in tank, immersed in solution, forming one or two complete rows (gaps filled with blank slides). Gels must be (just) covered. Leave 40 min.

8. Electrophoresis

Electrophoresis is performed for 30 min at 25 V (constant voltage setting). If there is too much electrolyte covering the slides, the current may be so high that it exceeds the maximum – so set this at a high level. If necessary, i.e. if 25 V is not reached, remove some solution. A current of 300 mA is normal. Perform electrophoresis in refrigerator or cold room.

(*Note:* The alkaline solution can be used several times – but it should be poured from the tank and mixed first.)

9. Neutralization

Give three washes (5 min each with neutralizing buffer in staining jar at 4°C.

10. Staining

Stain with 4′6-diamidine-2-phenylindol dihydrochloride (DAPI). Place 20 μl of a 1 μg/ml solution on to each gel and cover with a 22 × 22 mm coverslip. Keep slides in a dark, moist chamber until they are viewed. They may be left overnight before viewing, either stained or unstained (however, if stained, some fluorescence is lost).

Alternative stains

Propidium iodide (2.5 μg/ml), Hoechst 33258 (0.5 μg/ml), or ethidium bromide (20 μg/ml) can be used in place of DAPI for the visualization of comet DNA.

11. Quantitation

Computer image analysis

Several companies have created software which, linked to a closed-circuit digital camera mounted on the microscope, automatically analyses individual comet images. These programs are designed to differentiate the comet head from the tail, and to measure a variety of parameters including: tail length; percentage of total fluorescence in head and tail; and 'tail moment' (calculated in different ways but essentially representing the product of tail length and relative tail intensity).

- *% DNA in tail* is linearly related to DNA break frequency up to about 80% in tail, and this defines the useful range of the assay.
- *Tail length* tends to increase rapidly with dose at low levels of damage, but soon reaches its maximum. It is therefore the most sensitive parameter at near-background levels of damage.
- *Tail moment* is an attempt to combine the information of tail length and tail intensity, but it suffers from lack of linearity (Figure 7.1).

Visual analysis

It is perfectly possible to analyse comets quantitatively without image analysis software. The human eye can very rapidly discriminate comets representing different levels of damage, and we have developed a scheme for visual scoring based on five recognizable classes of comet, from class 0 (undamaged, no discernible tail) to class 4 (almost all DNA in tail, insignificant head) (Figure 7.2). Normally 100 comets are selected at random from each slide (avoiding the edges of the gel, where anomalously high levels of damage

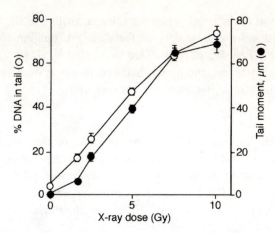

Figure 7.1 X-ray calibration of comet assay. Lymphocytes embedded in agarose were X-irradiated before lysis and electrophoresis. Fifty comets were analysed (from each of duplicate slides); mean % DNA in tail and mean tail moment are shown. X-rays induce 0.31 breaks per 10^9 daltons/Gy. (Redrawn from Collins *et al.*, 1996).

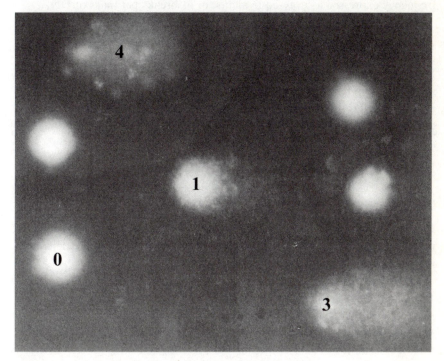

Figure 7.2 Examples of comets from human lymphocytes showing different degrees of damage (categorized in classes, 0–4, as shown).

are often seen). If each comet is given a value according to the class it is put into, an overall score can be derived for each gel, ranging from 0 to 400 arbitrary units. When slides are analysed in parallel by visual scoring and by computer image analysis, the match between results is excellent (Collins *et al.*, 1997). With practice, visual scoring is very quick.

Calculation
The control gels (no enzyme treatment) provide an estimate of the background of DNA strand breaks (SB). The enzyme-treated gels reveal strand breaks *and* oxidized bases (SB + OX). Assuming linearity of dose response, whether working in % DNA in tail or in arbitrary units, subtraction of (SB) from (SB + OX) gives a measure of oxidized pyrimidines/altered purines.

Calibration
Ionizing radiation produces strand breaks in DNA with known efficiency. If the breaks introduced in cells by different doses of X-rays are detected with the comet assay, a standard curve can be drawn (Figure 7.1), with break frequency expressed as gray-equivalents, or as breaks per mass of DNA.

12. Storage and re-examination

Place slides in a cool oven ($< 37°C$) until the gel has dried. Slides can then be stored at room temperature. For re-examination, stain as above. *Optionally*, add $100 \mu l$ of 1% standard agarose on top of the original layers, cover with coverslip, leave to set, remove coverslip and stain as before.

NOTES

(A) Lymphocytes: bulk preparation and storage

Lymphocytes can be kept frozen for several months without appreciable increase in background strand breakage. This is useful if large-scale experiments or complex human trials are carried out.

Required
- Blood: 10 ml in Li-heparin vacutainers
- Medium: RPMI 1640 (no FCS or other addition) at room temperature
- Lymphoprep from Nycomed; at room temperature
- Freezing medium: 90% FCS, 10% DMSO
- Sterile conditions

Procedure

1. Mix blood with 10 ml of RPMI.
2. Layer on 20 ml Lymphoprep in a 50 ml conical plastic centrifuge tube.
3. Spin 30 min at $700 \times g$, room temperature, no brake on centrifuge.
4. Remove buffy lymphocyte band from just above Lymphoprep using disposable 1 ml Pasteur pipette (Sterilin); collect about 10 ml and transfer to a new 50 ml plastic centrifuge tube.
5. Add RPMI to final volume of 50 ml. Spin 20 min at $700 \times g$, room temperature.
6. Decant supernatant, add fresh RPMI to volume of 25 ml, resuspend cells, count sample.
7. Spin 15 min at $700 \times g$, room temperature.
8. Decant supernatant; resuspend pellets in freezing medium at $3 \times 10^6/$ml and transfer to labelled cryotubes.
9. *Either* freeze using a programmed freezer at rate of $-1°/$min between $+4°C$ and $-30°C$, and then at rate of $-2°/$min between $-30°C$ and $-70°C$. Then transfer to liquid nitrogen.
10. *Or* place cryotubes in a thick-walled box of expanded polystyrene in a $-70°C$ or $-80°C$ freezer (the insulation ensures slow freezing). After 1 day, transfer to liquid nitrogen. (We find that the second of these methods is more successful.)
11. Thaw cells quickly (e.g. by holding tube in a $37°C$ water bath), dilute them into several millilitres of appropriate growth medium, spin as above (15 min) and resuspend the pellet in medium.
12. Proceed with appropriate treatment prior to comet assay.

(B) Enzymes

Endonuclease III (endo III) and formamidopyrimidine glycosylase (fpg) are isolated from bacteria containing overproducing plasmids. Because such a high proportion of protein is the enzyme, a crude extract is perfectly satisfactory; in our experience there is no non-specific nuclease activity. The enzyme extracts are best obtained from a laboratory producing them, although there are commercial sources. On receipt, the enzyme (which should have been refrigerated in transit) should be dispensed into small aliquots (say, $2 \mu l$) and stored at $-80°C$. This minimizes repeated freezing and thawing.

The final dilution of the working solution will vary from batch to batch; the supplier should suggest what this should be, but it is best to check personally, as described below.

Examples

(1)   If endo III has a suggested dilution of 1000 ×, dilute one 2 μl aliquot with buffer to 2 ml and store in suitable aliquots (e.g. 300 μl – enough for six gels) at −80°C.

(2)   As fpg is less stable than endo III, repeated freezing/thawing must be minimized. If the suggested dilution is 3000 ×, first dilute one 2 μl aliquot to 200 μl (100 ×) and store 10 μl aliquots at −80°C. *For this dilution, use buffer containing 10% glycerol.* For use, dilute one 10 μl aliquot to 300 μl with buffer (no glycerol) and use at once; do not refreeze.

Checking final dilution

We use, as substrate, HeLa cells treated with H_2O_2 (to induce oxidative damage – strand breaks and oxidized bases) and incubated for 1 h (long enough for repair of strand breaks but not of oxidized bases). As a negative control, HeLa cells without H_2O_2 treatment have virtually no oxidized bases of endogenous origin, and so they provide a good test for the lack of non-specific nuclease activity in the extract. Embed the test cells in agarose and lyse as usual; wash with enzyme buffer; and incubate with serial dilutions of enzyme in buffer (for example, 1/300, 1/1000, 1/3000, 1/10 000, 1/30 000). It will probably be found that the higher concentrations of enzyme *do* produce some non-specific breaks in the untreated control nucleoids, and the lowest concentrations produce no breaks at all above background even in the nucleoids from the treated cells. There should be a range of concentrations that give no breaks in the control nucleoids and a significant number of breaks in the treated nucleoids. Choose from this range the concentration that gives the highest yield of breaks in treated nucleoids. Then test this concentration and one on each side of it on normal lymphocytes (not treated with H_2O_2) to check that endogenous damage is detected optimally at the same concentration of enzyme.

(C) Treatment with H_2O_2

For example, to look at antioxidant resistance, or capacity for DNA repair:

1.   Stock H_2O_2 is 8.8 M.
2.   Dilute 11.5 μl in 1 ml H_2O (0.1 M)
3.   Keep this as working stock for 1 week at most and then discard
4.   Dilute 10 μl of 0.1 M solution in 1 ml PBS (1 mM)
5.   Further dilutions in PBS, to required concentrations for experiment (e.g. 25 μM = 975 μl PBS + 25 μl of 1 mM solution).

6. Add to cells in tube or on dish. Leave on ice for 5 min. Collect cells (trypsinize if necessary, centrifuge, wash with PBS), embed, etc. as above. *Alternatively*: after cells are embedded on agarose, add 50 μl of required concentration of H_2O_2, cover with 22 × 22 coverslip, leave in cold for 5 min. Wash with PBS before lysis.

When H_2O_2 has been used, it is advisable to isolate any control slides by placing in separate vessels during lysis; otherwise strand breaks may occur in the control slides.

(D) Measuring antioxidant resistance

Cells are challenged *in vitro* with H_2O_2 and the yield of strand breaks, measured with the comet assay, indicates the antioxidant status of the cells. Resistance to H_2O_2 is increased by supplementation with dietary antioxidants (Duthie *et al.*, 1996).

(E) Monitoring recovery from oxidative damage *in vitro*

After a challenge with H_2O_2, cells are incubated in suitable medium at 37°C and the removal of damage by cellular repair can, in theory, be monitored. Removal of breaks from lymphocyte DNA tends to be very slow (and removal of oxidized bases even slower) (Collins *et al.*, 1997); it seems that the apparent slow repair is actually the result of a continuing input of new oxidative damage occurring in the high oxygen tension of the culture incubator (compared with that obtaining in the body).

REFERENCES

Brooks, R.A. and Winton, D.J. (1996) Determination of spatial patterns of DNA damage and repair in intestinal crypts by multi-cell gel electrophoresis. *J. Cell Sci.*, **109**, 2061–2068.

Collins, A.R., Duthie, S.J. and Dobson, V.L. (1993) Direct enzymic detection of endogenous oxidative base damage in human lymphocyte DNA. *Carcinogenesis*, **14**, 1733–1735.

Collins, A.R., Dušinská, M., Gedik, C.M. and Štětina, R. (1996) Oxidative damage to DNA: do we have a reliable biomarker? *Env. Health Perspect.* **104**, Suppl. 3, 465–469.

Collins, A., Dušinská, M., Franklin, M., Somorovská, M., Petrovská, H., Duthie, S., Fillion, L., Panayiotidis, M., Rašlová K, and Vaughan, N. (1997) Comet assay in human biomonitoring studies: reliability, validation, and applications. *Env. Molec. Mutagenesis*, **30**, 139–146.

Dušinská, M. and Collins, A. (1996) Detection of oxidised purines and UV-induced photoproducts in DNA of single cells, by inclusion of lesion-specific enzymes in the comet assay. *ATLA* **24**, 405–411.

Duthie, S.J., Ma, A., Ross, M.A. and Collins, A.R. (1996) Antioxidant supplementation decreases oxidative DNA damage in human lymphocytes. *Cancer Res.*, **56**, 1291–1295.

Singh, N.P., McCoy, M.T., Tice, R.R. and Schneider, E.L. (1988) A simple technique for quantitation of low levels of DNA damage in individual cells. *Exp. Cell Res.*, **175**, 184–191.

8 Measuring Oxidative DNA Damage by Alkaline Elution

Michael Pflaum and Bernd Epe

INTRODUCTION

Oxidative DNA modifications, which include single-strand breaks (SSB), sites of base loss (AP sites) and many types of base modifications, are generated in cells not only under the influence of various xenobiotics (oxidants), solar light or ionizing radiation, but also under normal growth conditions. In the latter case, reactive oxygen species generated in the oxygen metabolism of the cells are assumed to be responsible. The balance between the generation of the oxidative lesions and their removal by repair enzymes results in background (steady-state) levels that can be observed in all types of cells.

For better assessment of the role of oxidative DNA modifications in mutagenesis, carcinogenesis and various degenerative diseases, sensitive quantification methods have been developed. These include the quantification of 8-hydroxyguanine (8-oxoG) by HPLC with an electrochemical detector (HPLC–ECD) (Kasai *et al.*, 1986) and the determination of various base modifications by gas chromatography coupled with mass spectrometry (GCMS) (Nackerdien *et al.*, 1992). In DNA from mammalian cells, SSB can be measured very sensitively by several techniques, including alkaline elution, DNA unwinding and single cell gel electrophoresis. The same techniques can also be used to quantify AP sites and certain base modifications if purified repair endonucleases are used as probes (Fornace, 1982; Epe and Hegler, 1994; Collins *et al.*, 1993; Hegler *et al.*, 1993; Hartwig *et al.*, 1996). The enzymes incise the DNA at their substrate modifications, converting them into SSB. Therefore, if cellular DNA is incubated with a repair endonuclease

Measuring in vivo *Oxidative Damage: A Practical Approach.* Edited by J. Lunec and H. R. Griffiths. © 2000 by John Wiley & Sons, Ltd. ISBN 0 471 81848 8.

prior to analysis by one of the above techniques, the sum of already existing SSB and endonuclease-sensitive site (ESS) modifications is obtained. Some repair endonucleases suitable for the analysis of oxidative DNA modifications are listed in Table 8.1. With the alkaline elution technique described below, the detection limit is approximately 0.05 modifications per 10^6 base pairs in 10^6 cells (Epe and Hegler, 1994; Pflaum *et al.*, 1997).

Table 8.1 Some repair endonucleases suitable for analysis of oxidative DNA damage

| Repair endonuclease | Recognition spectrum[a] | | | |
| | Sites of base loss (AP sites) | | | |
	Regular[b]	1′-ox.[c]	4′-ox.[d]	Base modifications
Fpg protein	+	−	+	8-oxoG[e], Fapy[f]
Endonuclease III	+	−	+	5,6-dihydropyrimidines, hyd[g]
T4 Endonuclease V	+	−	(+)[h]	CPD[i]
Endonuclease IV	+	+	+	−
Exonuclease III	+	+	(+)[h]	−

[a] See Boiteux, 1993; Demple and Harrison, 1994; Karakaya *et al*, 1997.
[b] Unmodified deoxyribose moiety.
[c] Deoxyribose modified in the 1′ position.
[d] Deoxyribose modified in the 4′ position.
[e] 7,8-Dihydro-8-oxoguanine (8-hydroxyguanine).
[f] Formamidopyrimidines (imidazole ring-opened purines).
[g] 5-Hydroxy-5-methylhydantoin and other ring-contracted and fragmented pyrimidines.
[h] Recognition requires high enzyme concentrations.
[i] Cyclobutane pyrimidine dimers.

PRINCIPLE OF THE ASSAY

The alkaline elution technique was originally developed by Kohn *et al.* (1976). Suspended cells are lysed on a membrane filter and all cell constituents other than DNA are eluted from the filter with the lysis solution. The DNA remaining on the filter is then eluted under alkaline conditions (pH 12.2), which cause DNA strand separation. Large fragments of single-stranded DNA (few breaks) elute slower than small fragments (many breaks). Within a certain range, the DNA elution rate is proportional to the number of SSB in the DNA. If the DNA on the filter is incubated with a repair endonuclease prior to alkaline elution, the elution rate reflects the sum of SSB and ESS.

SCOPE AND LIMITATIONS

Major characteristics of the assay are as follows.

1. Most repair endonucleases are not specific for a single substrate, but recognize several modifications (see Table 8.1). Obviously, the recognition of additional (unknown) modifications is difficult to exclude. More specific information can often be obtained when several repair endonucleases (with differing substrate specificities) are used in parallel.
2. Not only base modifications but also SSB and sites of base loss (AP sites) can be quantified.
3. Due to the alkaline conditions used in the assay, some very alkali-labile modifications may be detected as SSB.
4. The sensitivity of the assay is very high in comparison with that of other techniques such as HPLC–ECD and GCMS. On the other hand, high levels of damage (> 5 modifications per 10^6 base pairs) can not be quantified.
5. The method does not involve DNA hydrolysis and therefore avoids this major source of artefacts.
6. The method is not applicable for mitochondrial DNA. A relaxation assay in combination with repair endonucleases can be used instead (Hegler *et al.*, 1993).

REAGENTS

- Phosphate buffered saline with glucose (PBSG), sterile filtered

  – NaCl: 8.0 g/l
  – KCl: 0.2 g/l
  – $Na_2HPO_4 \times 2H_2O$: 1.15 g/l
  – KH_2PO_4: 0.2 g/l
  – $CaCl_2 \times 2H_2O$: 0.135 g/l
  – $MgCl_2 \times 6H_2O$: 0.1 g/l
  – glucose: 1.0 g/l

- Phosphate buffered saline, free of calcium and magnesium (PBSCMF), autoclaved

  – NaCl: 8.0 g/l
  – KCl: 0.2 g/l
  – $Na_2HPO_4 \times 2H_2O$: 1.15 g/l
  – KH_2PO_4: 0.2 g/l

- BE1 buffer, pH 7.5, autoclaved

  – Tris/HCl: 20 mM
  – NaCl: 100 mM
  – Na$_2$-EDTA: 1 mM

- BE15 buffer, pH 7.5, autoclaved

  – Tris/HCl: 20 mM
  – NaCl: 100 mM
  – Na$_2$-EDTA: 15 mM

- TC buffer, pH 8.0, autoclaved

  – Tris/HCl: 50 mM
  – CaCl$_2$: 25 mM

- Lysis buffer, pH 10.0

  – glycine: 100 mM
  – Na$_2$-EDTA: 20 mM
  – SDS: 20 g/l

- Cleaning buffer, pH 10.0

  – Na$_2$-EDTA: 20 mM

- Elution buffer, pH 12.2

  – H$_4$-EDTA (free acid !): 20 mM
  – tetraethylammonium hydroxide: as required to adjust the pH to 12.2

- Bisbenzimide stock solution (Hoechst No. 33258): 0.15 mM in H$_2$O
- Phosphate buffer I, pH 6.0, autoclaved

  – Na-phosphate: 100 mM

- Phosphate buffer II, pH 7.2, autoclaved, with bisbenzimide

  – Na-phosphate: 100 mM
  – bisbenzimide stock solution: 1% (v/v)

INSTRUMENTS

- Swinnex filter holder (25 mm diameter)
- Polycarbonate membrane filters (25 mm diameter, 2 μm pore size)
- Multichannel peristaltic pump with high precision speed control

- Water bath
- Multichannel fraction collector
- Thermostat (4°C to 37°C)
- Fluorescence spectrometer

The set-up of the alkaline elution system in shown in Figure 8.1.

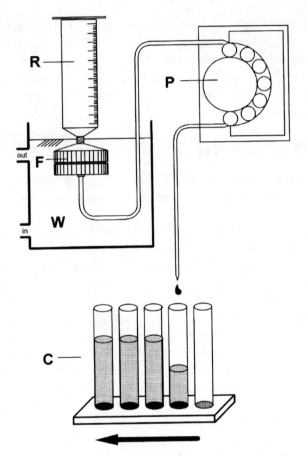

Figure 8.1 Alkaline elution system with buffer reservoir (R), filter holder with membrane filter (F), water bath (W), multi-channel peristaltic pump (P) and fraction collector (C). Only one channel is shown.

REPAIR ENDONUCLEASES

Some (but not all) repair endonucleases listed in Table 8.1 are commercially available. Prior to use, the enzyme activity of each new lot of enzyme preparation has to be tested. This can be done in a relaxation assay with supercoiled DNA containing suitable reference modifications, as described elsewhere (Epe and Hegler, 1994). Saturation curves (enzyme concentration dependencies) for the incision reaction should be measured. As a rule, saturating enzyme concentrations in the cell-free relaxation assay are similar to those required for complete incision in the alkaline elution assay.

Frozen aliquots of repair endonucleases in sterile BE1 buffer may be kept for months at $-70°$, but repeated freezing/thawing can drastically reduce enzyme activities. Incubation with exonuclease III, which requires Ca^{2+} or Mg^{2+}, is carried out in TC buffer supplemented with $100\,\mu g/ml$ bovine serum albumine (BSA). All the other endonucleases in Table 8.1 are used in BE1 buffer with $100\mu g/ml$ BSA.

PROBLEMS AND PITFALLS

- Bisbenzimide is photolabile. The stock solution should be added to the phosphate buffer immediately before application and stored in a brown glass bottle.
- In order to avoid repair of SSB and (much less a problem) DNA base modifications, cells must be kept on ice until lysis.
- The lysis step is very delicate. Lysis buffer contains SDS, which precipitates at temperatures less than 15°C. This conflicts with the requirement of keeping cells cold prior to lysis.
- It is very important to withdraw any traces of lysis buffer prior to incubation of DNA with repair enzymes. During washing steps with BE1 buffer, the inner walls of the reservoir must be carefully rinsed. If the pump speed is not exactly the same in all channels, the pump is stopped when the surface of BE1 buffer reaches the bottom of the syringe of the fastest channel. Residual BE1 buffer in the other reservoirs is removed with a pipette.
- The inclusion of one filter that is not loaded with cells, but is otherwise treated in the same way as the other filters, is recommended. The fluorescence data of this filter serve as zero values for the corresponding DNA-containing solutions.
- Provided that SSB and ESS are randomly distributed within the DNA, the elution profile obtained when the logarithm of the percentage of DNA still on the filter is plotted against the elution time should be a straight line. Inaccurate measurements of the low amounts of DNA in the late fractions and also the presence of alkali-labile DNA modifications

and several other factors can give rise to convex or concave curves, in particular close to the end of the elution. Only the linear range of the semi-logarithmic graph may be used for calculating the numbers of modifications.

- Not only too low but also excessively high concentrations of repair endonucleases should be avoided, to ensure that only substrate modifications are incised and that the activities of any contaminating endonucleases are negligible. Experiments with various enzyme concentrations should be carried out to test for complete incision.

PROCEDURE

1. Equip Swinnex-type filter holders, filled with water, with polycarbonate filters (25 mm diameter) and connect with a 25 ml polyethylene Luer-lok syringe serving as reservoir for the solutions to be pumped through the filter (Figure 8.1).

2. Fix the filter devices to a suitable rack, connected with a multi-channel peristaltic pump, and replace the water by PBSG: load the reservoir with 5 ml PBSG, pumped into the system with maximum pump speed (60 ml/h). The system must not run dry at any time of the procedure before step 14. Air bubbles have to be withdrawn from the filter device.

3. Cool the filters containing PBSG down to 4°C by placing the whole rack into a water bath.

4. Load each filter with 10^6 cells at maximum pump speed.

5. Wash the cells twice with 3 ml of ice-cold PBSCMF each.

6. Remove the water bath. Without delay, add 2 ml lysis buffer (room temperature) and pump this into the system at maximum pump speed.

7. Lyse the cells with 5 ml lysis buffer at 25°C (water bath) within 1 h (pump speed 5 ml/h).

8. Completely wash out lysis buffer by rinsing five times with 5 ml buffer BE1 each, at maximum pump speed. During washing steps, rise the temperature of the water bath to 37°C.

9. Apply 2 ml repair endonuclease solution and pump through the filter at 37°C. Pump the first 1 ml with maximum speed, the second within 30 min. To measure SSB only, this step is carried out without endonucleases.

10. After washing with 5 ml buffer BE1, pass 5 ml lysis buffer containing 500 μg/ml proteinase K through the filter at 25°C within 30 min.

11. A second washing procedure follows, as described in step 8.
12. Add 5 ml cleaning buffer and pump with maximum pump speed.
13. Load the reservoirs with 25 ml elution buffer. Elute the DNA with 2.3 ml/h and collect for 10 h in 4.6 ml fractions.
14. Mix each fraction with 4.6 ml phosphate buffer I, and after 5 min with 4.6 ml phosphate buffer II containing bisbenzimide. Measure the fluorescence of each solution at excitation wavelength 360 nm and emission wavelength 450 nm.
15. The amount of the DNA remaining on the filter after 10 h also has to be determined. Put the polycarbonate filters and the sieves of the filter holders into a suitable beaker. Cover with 10 ml of elution buffer (including the elution buffer that is still in the elution system at the end of the elution) and shake for 2 h at 60°C in a water bath. Treat the aliquots of the resulting DNA solutions as described in step 14.

CALCULATION OF ELUTION RATES (FOR EACH FILTER)

The amounts of DNA in each of the fractions and on the filter after 10 h elution time are calculated as follows:

$$DNA_i = F \times V$$

where DNA_i = amount of DNA in fraction i, in arbitrary units; F = relative fluorescence of the DNA solution (excitation wavelength 360 nm, emission wavelength 450 nm) in arbitrary units; and V = volume of fractions (4.6 ml) or of the dissolved filter DNA after 10 h elution time (10.0 ml).

The portion of DNA remaining on the filter after elution of fraction n ($n = 1$ to 5) is then calculated as follows:

$$FRAC_n = \left(1 - \frac{\sum_{i=1}^{n} DNA_i}{\sum_{i=1}^{5} DNA_i + DNA_F}\right) \times 100$$

where DNA_i = amount of DNA in fraction i; DNA_F = amount of DNA on the filter after elution of the last fraction (fraction 5); and $FRAC_n$ percentage of DNA on the filter after elution of the nth fraction (% of total DNA).

The logarithms of the $FRAC_n$ values are plotted against the elution time. Straight lines should be obtained. The rate of the alkaline elution is then calculated:

$$m = \log(FRAC_1) - \log(FRAC_5)$$

where m = elution rate; and $FRAC_1$, $FRAC_5$ = percentage of DNA on the filter after elution of the first and fifth fraction, respectively.

Calculation of single-stand breaks (SSB) and endonuclease-sensitive sites (ESS)

When a repair endonuclease is applied (step 9), the elution rates m correspond to the sum of SSB and ESS. Values m_{ESS} that correspond to the number of ESS are obtained by subtraction of the m_{SSB} values obtained from filters without repair endonuclease treatment.

Except for the determination of steady-state levels of oxidative modifications, elution rates observed with untreated control cells are subtracted from the values observed with treated cells.

In order to calculate the absolute number of SSB and ESS, the elution rates are compared with the elution rate observed with cells exposed to a defined dose of -irradiation at $0°C$: 3 Gy induces 0.54 SSB per 10^6 base pairs (Kohn et al. 1976).

$$N_{mod} = m \times \frac{0.54}{m_{3Gy}} \times 10^{-6}$$

where N_{mod} = number of SSB or ESS per base pair corresponding to the elution rate m; m_{3Gy} = elution rate observed with cells exposed to 3 Gy gamma-radiation at $0°C$ after subtraction of the elution rate observed with untreated cells; and m = elution rate.

Detection range

SSB generated by various doses of ionizing radiation are also used to determine the range in which the elution rate m increases linearly with the number of SSB and ESS. For the conditions described above, a linear dose response was observed between 0.3 and 9.0 Gy, equivalent to 0.05–1.6 modifications per 10^6 base pairs. To measure higher levels of damage, the pump speed during the alkaline elution can be increased. With a pump speed of

0.5 ml/min, damage levels between 0.2 and 6.0 modifications per 10^6 base pairs can be determined.

REFERENCES

Boiteux, S. (1993) Properties and biological functions of the NTH and FPG proteins of *Escherichia coli*: two DNA glycosylases that repair oxidative damage in DNA. *Photochem. Photobiol. B.*, **19**, 87–96.

Collins, A.R., Duthie, S.J. and Dobson, V.L. (1993) Direct enzymatic detection of endogenous oxidative base damage in human lymphocyte DNA. *Carcinogenesis*, **14**, 1733–1735.

Demple, B. and Harrison, L. (1994), Repair of oxidative damage to DNA: enzymology and biology. *Annu. Rev. Biochem.*, **63**, 915–948.

Epe, B. and Hegler, J. (1994), Oxidative DNA damage: endonuclease fingerprinting. *Methods Enzymol.*, **234**, 122–131.

Fornace, Jr, A. J. (1982) Measurement of *M. luteus* endonuclease-sensitive lesions by alkaline elution. *Mutat. Res.*, **94**, 263–276.

Häring, M., Rüdiger, H., Demple, B., Boiteux, S. and Epe, B. (1994), Recognition of oxidized abasic sites by repair endonucleases. *Nucleic Acids Res.*, **22**, 2010–2015.

Hartwig, A., Dally, H. and Schlepegrell, R. (1996) Sensitive analysis of oxidative DNA damage in mammalian cells: use of the bacterial Fpg protein in combination with alkaline unwinding. *Toxicology*, **110**, 1–6.

Hegler, J., Bittner, D., Boiteux, S. and Epe, B. (1993) Quantification of oxidative DNA modifications in mitochondria. *Carcinogenesis*, **14**, 2309–2312.

Karakaya, A., Jaruga, P., Bohr, V.A., Grollman, A.P. and Dizdaroglu, M. (1997) Kinetics of excision of purine lesions from DNA by *Escherichia coli* Fpg protein. *Nucleic Acids Res.*, **25**, 474–479.

Kasai, H., Crain, P.F., Kuchino, Y., Nishimura, S., Ootsuyama, A. and Tanooka, H. (1986) Formation of 8-hydroxyguanine moiety in cellular DNA by agents producing oxygen radicals and evidence for its repair. *Carcinogenesis*, **7**, 1849–1851.

Kohn, K.W., Erickson, L.C., Ewig, R.A.G. and Friedman, C.A. (1976) Fractionation of DNA from mammalian cells by alkaline elution. *Biochemistry*, **15**, 4629–4637.

Nackerdien, Z., Olinski, R. and Dizdaroglu, M. (1992) DNA base damage in chromatin of γ-irradiated cultured human cells. *Free Radical Res. Commun.*, **16**, 259–273.

Pflaum, M., Will, O. and Epe, B. (1997) Determination of steady-state levels of oxidative DNA base modifications in mammalian cells by means of repair endonucleases. *Carcinogenesis*, **18**, 2225–2231.

Part IV Molecular-based Assays

9 A ^{32}P-postlabelling Protocol to Measure Oxidative DNA Damage

George D.D. Jones and Michael Weinfeld

INTRODUCTION

The ^{32}P-postlabelling assay is widely employed for the measurement of carcinogen–DNA adducts and several other types of lesions in DNA, including oxidative damage (Phillips, 1997). In the original assay described by Randerath *et al.* (1981), the DNA is first digested by micrococcal nuclease (MN) and calf spleen phosphodiesterase (CSPD) to yield nucleoside 3′-monophosphates (both normal and modified), which are then 5′-^{32}P end-labelled by incubation with [γ-^{32}P]ATP and T4 polynucleotide kinase (T4PNK). The radiolabelled compounds are then separated by either multi-dimensional thin layer chromatogrpahy (TLC) or high-performance liquid chromatography with radioactivity detection (^{32}P-HPLC) (Möller *et al.*, 1993).

The assay has two important advantages. Firstly, there is no requirement for prelabelling the DNA, which makes the assay useful for the study of DNA isolated from tissues. Secondly, because of the availability of [γ-^{32}P]ATP of high specific activity, the assay permits detection at the femtomole level. There are, however, several major drawbacks, particularly when assaying for oxidative DNA damage: (i) the polynucleotide kinase must be able to act on modified nucleoside 3′-monophosphates; (ii) there is the possibility of additional adventitious oxidative damage formation during the labelling stage, particularly from the radiolytic decay of ^{32}P; (iii) the labelled modified nucleoside bisphosphates must be separable from the high background of normal nucleoside bisphosphates; and (iv) common oxidative lesions that involve base loss, such

Measuring in vivo *Oxidative Damage: A Practical Approach.* Edited by J. Lunec and H. R. Griffiths. © 2000 by John Wiley & Sons, Ltd. ISBN 0 471 81848 8.

as abasic sites and deoxyribose fragments, are not detected by this approach.

The problems of poor labelling efficiency, radiolytic autooxidation, and poor chromatographic resolution of products are well illustrated in the reports of efforts to detect thymine glycols and 8-oxoguanine, well-known oxidative base lesions, in DNA (Hegi *et al.*, 1989; Reddy *et al.*, 1991; Povey *et al.*, 1993; Möller and Hofer, 1997; Podmore *et al.*, 1997; Schuler *et al.*, 1997). To alleviate these problems, it was found necessary to introduce additional lesion enrichment procedures to improve assay sensitivity.

In our respective laboratories we have been developing and exploiting an alternative postlabelling approach that overcomes the problems noted above. The assay is based upon the observation that particular DNA lesions prevent hydrolysis by snake venom phosphodiesterase (SVPD) and DNase I of the adjacent 5′-internucleotide phosphodiester linkage (Stuart and Chambers, 1987; Liuzzi *et al.*, 1989; Weinfeld and Buchko, 1993; Weinfeld *et al.*, 1993). Therefore, digestion of damaged DNA with these enzymes and an alkaline phosphatase yields certain damage as lesion-containing 'dinucleoside' monophosphates (dNpX) where the damage (X) is 3′ to a normal nucleoside 3′-monophosphate moiety (dNp-). Because the nucleoside 3′-monophosphate moiety is unmodified, these dimers are ready substrates for polynucleotide kinase and [γ-^{32}P]ATP, allowing for their *exclusive* labelling and detection as 5′-^{32}P end-labelled dimer species (^{32}P-dNpX) (Figure 9.1). The undamaged bases are recovered as mononucleosides, which are not substrates for the kinase and so remain unlabelled. To consume the excess [γ-^{32}P]ATP (which would otherwise impede PAGE analysis of the labelled damage-containing dimers species; see below), oligo(dT)$_{16}$ is added along with a second aliquot of the kinase to transfer the excess ^{32}P label on to the 5′-end of the large oligonucleotide. Alternatively, apyrase is added to each sample to catalyse the hydrolysis of [γ-^{32}P]ATP, releasing the radiolabel as inorganic phosphate (^{32}Pi) (Figure 9.1).

The labelled 'dinucleotides' are easily separated and quantified by 20% denaturing polyacrylamide gel electrophoresis (PAGE) and reverse-phase HPLC, and absolute identification achieved by co-comparison (via PAGE and HPLC) with prepared marker compounds (cf. Weinfeld and Soderlind, 1991). The advantages of SVPD-postlabelling over other postlabelling methodologies are: (i) the absence of labelling of undamaged DNA (so there is no need for a damage enrichment step); (ii) the improved labelling of lesion-containing molecules (as it is an *un*modified dNp moiety that is labelled); (iii) the capacity to assay lesions undetectable by other approaches (or detected with very low efficiency), in particular thymine glycols and lesions that

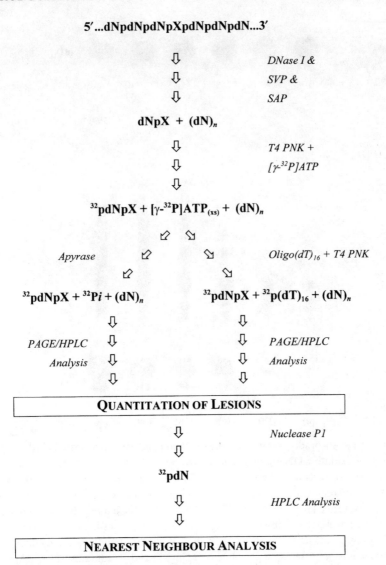

5′...dNpdNpdNpXpdNpdNpdN...3′

⇩   *DNase I &*
⇩   *SVP &*
⇩   *SAP*

dNpX + (dN)$_n$

⇩   *T4 PNK +*
⇩   *[γ-^{32}P]ATP*
⇩

^{32}**pdNpX + [γ-^{32}P]ATP$_{(xs)}$ + (dN)$_n$**

↖ ↘
Apyrase ↖ ↘ *Oligo(dT)$_{16}$ + T4 PNK*
↖ ↘

^{32}**pdNpX + ^{32}Pi + (dN)$_n$** ^{32}**pdNpX + ^{32}p(dT)$_{16}$ + (dN)$_n$**

⇩   ⇩
PAGE/HPLC ⇩   ⇩ *PAGE/HPLC*
Analysis ⇩   ⇩ *Analysis*
⇩   ⇩

QUANTITATION OF LESIONS

⇩   *Nuclease P1*
⇩

^{32}**pdN**

⇩   *HPLC Analysis*
⇩

NEAREST NEIGHBOUR ANALYSIS

Figure 9.1 Scheme of the postlabelling assay. Damaged DNA is digested to mono-nucleosides (dN) and 'dinucleoside' monophosphates (NpX) where X represents a damaged base or deoxyribose. The NpX species are radioactively phosphorylated (^{32}pNpX) and, after transferring most of the excess label to a large oligonucleotide, (dT)$_{16}$, or hydrolysing the excess [γ-^{32}P]ATP to release the radiolabel as inorganic phosphate (^{32}Pi), the ^{32}pNpX products can be analysed by polyacrylamide gel electrophoresis (PAGE) and/or HPLC. The labelled 5′-neighbouring nucleotide (^{32}pdN) can be released by nuclease P1-mediated cleavage of the phosphodiester bond in the purified ^{32}pNpX molecules and identified by HPLC.

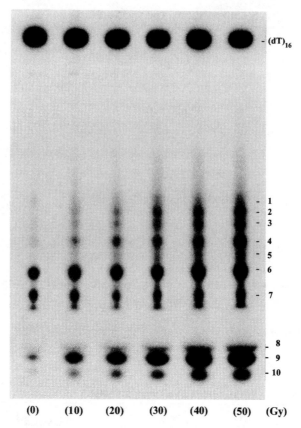

Figure 9.2 Autoradiograph of a polyacrylamide gel showing postlabelled products following oxidative DNA damage generated by γ-irradiation (0–50 Gy). Some of the detectable products have been identified (Weinfeld and Soderlind, 1991). Bands 1–3 contain the labelled thymine glycol-containing dinucleotides; ^{32}pGpT^g is in band 1, ^{32}pApT^g and ^{32}pTpT^g are in band 2, and ^{32}pCpT^g is in band 3. Bands 8–10 are the labelled phosphoglycolate species (^{32}pNpg), resulting from hydroxyl radical attack and oxidation at the C-4′ carbon atom in deoxyribose; band 8 contains the labelled deoxyguanosine 3′-phosphoglycolate (^{32}pGpg), band 9 contains the ^{32}pApg and ^{32}pTpg and band 10 contains ^{32}pCpg. The lesions in bands 4 and 5 have yet to be identified, but they can be removed from damaged DNA by *Escherichia coli* endonuclease IV. Bands 6 and 7 are seen with the untreated control. The labelled oligo(dT)$_{16}$ band, reflecting excess [γ-^{32}P]ATP remaining after the labelling of the damage-containing dimer species (see Figure 9.1), can be seen at the top of the gel. Not shown in this figure are three bands containing inorganic mono-, di- and triphosphate that migrate faster than the phosphoglycolate molecules. The apparatus employed was the Gibco BRL model S2 sequencing gel electrophoresis apparatus, using 0.8 mm spacers, as described in the Methods section.

involve base loss and deoxyribose fragmentation; and (iv) nullification of adventitous oxidative effects during labelling (Jones *et al.*, 1999).

This procedure has been used to quantify radiation-induced thymine glycol and phosphoglycolate termini (Weinfeld and Soderlind, 1991), to compare oxidative lesions produced by Fenton chemistry and ionizing radiation (Weinfeld *et al.*, 1996; Jones and Weinfeld, unpublished results) and to detect phosphoglycolate lesions produced in cellular DNA by oxidative stress (Bertoncini and Meneghini, 1995). The assay has also been used to examine the influence of nitrogen, oxygen and nitroimidazole radiosensitizers on radiogenic DNA damage (Buchko and Weinfeld, 1993), to examine oxidative damage induced by the bioreductive drug Tirapazamine (Jones and Weinfeld, 1996) and to quantify abasic sites (Weinfeld *et al.*, 1990).

METHODS

Materials

Reagents
- Oligo $(dT)_{16}$, from Pharmacia Biotech (Catalogue No. 27–7609–01) and dissolved in distilled water to a concentration of $5\,AU_{260}$ units/ml.
- Calf thymus DNA (type 1, sodium salt), from Sigma Chemical Co. (Catalogue No. D1501); salmon testes DNA (sodium salt), from Sigma Chemical Co. (Catalogue No. D1626).
- $[\gamma\text{-}^{32}P]ATP$ (*Redi*vue™, $\sim$3000 Ci/mmol), from Amersham International plc (Catalogue No. AA0068).
- Ultra pure ACCUGEL™ 19:1, sequencing grade 40% acrylamide:bisacrylamide solution (gas stabilized), purchased from National Diagnostics (Catalogue No. EC-850). (Care must be taken in handling this solution because acrylamide is a neurotoxin and suspected carcinogen.)
- Ammonium persulfate (*UltraPure*™), from Gibco BRL (Catalogue No. 15523–012) – prepared as a 3% w/v solution in distilled water. The solution can be kept for up to 2 weeks at 4°C.

Buffers
- $10 \times$ TE: $100\,\text{mM}$ Tris-HCl (pH 7.4), $10\,\text{mM}$ EDTA (pH 8.0)
- DNA digestion buffer: 1X TE plus $4\,\text{mM}$ $MgCl_2$
- $10 \times$ T4 Polynucleotide kinase reaction buffer (from United States Biochemical (USB), Catalogue No. 70083): $0.5\,\text{M}$ Tris-HCl (pH 7.6), $100\,\text{mM}$ $MgCl_2$, $100\,\text{mM}$ 2-mercaptoethanol; $10 \times$ T4 polynucleotide kinase reaction buffer (from Promega, Catalogue No. C131A): $0.7\,\text{M}$ Tris-HCl

(pH 7.6), 100 mM MgCl$_2$, 50 mM DTT. These buffer concentrates are supplied along with the T4PNK enzymes by the respective suppliers.

- 10 × TBE Ultra pure buffer concentrate, 0.89 M Tris Borate (pH8.3), 20 mM EDTA, electrophoretic sequencing grade: from National Diagnostics (Catalogue No. EC-860).
- Gel loading buffer: 90% formamide, 0.02% bromophenol blue, and 0.02% xylene cyanol in 1 × TBE.
- Nuclease P1 buffer: 10 mM sodium acetate (pH 5.3), 1 mM ZnCl$_2$.

Enzymes

- Deoxyribonuclease I (Type II from bovine pancreas), from Sigma (Catalogue No. D4527), dissolved in distilled water to a concentration of 40 units/μl, aliquoted and stored at −20°C.
- Phosphodiesterase I (Type IV from *Crotalus atrox*), from Sigma (Catalogue No. P4506), resuspended in distilled water to a concentration of 0.011 units/μl, aliquoted and stored at −20°C. (Care should be taken in handling this material since it is a relatively crude preparation of snake venom.)
- Shrimp alkaline phosphatase (1 unit/μl): purchased from USB (Catalogue No. 70092).
- T4 polynucleotide kinase (30 units/μl): purchased from USB (Catalogue No. 70031); and T4 polynucleotide kinase (5–10 units/μl): purchased from Promega (Catalogue No. M4103). Both suppliers provide their enzymes with an appropriate 10 × kinase reaction buffer (see above).
- Apyrase (Grade VIII from potato), from Sigma (Catalogue No. A6160), dissolved in distilled water to a concentration of 0.1 units/μl, aliquoted and stored at −20°C.
- Nuclease P1, from Pharmacia Biotech (Catalogue No. 27–0852–01), dissolved at 1 mg/ml (~600 units/ml) in 10 mM sodium acetate (pH 5.3), aliquoted and stored at −20°C.

Definitions of the units of these enzymes are those given by the suppliers.

Preparation of polyacrylamide gels

Apparatus

We employ a power supply, capable of delivering at least 1000 V, and have traditionally used a Hoefer SE 620 electrophoresis unit (capable of running two gels simultaneously) using glass plates 32 cm long, 1.5 mm spacers and a 20-tooth comb (see below). Unfortunately, the SE 620 unit was discontinued by Hoefer, and subsequently Hoefer Scientific Instruments was acquired by Pharmacia Biotech, who continue to supply Hoefer equipment but not the SE 620 unit. It is possible to adapt one of the shorter SE 600 units (Pharmacia Biotech, Catalogue No. 80–6171–58), which uses 16 cm plates, to accommodate the longer 32 cm plates by constructing a deeper (41 cm) lower buffer chamber (fitted with a platinum electrode) and by obtaining the following additional items from Pharmacia Biotech: 18 cm (width) × 32 cm (length) glass plates (Catalogue No. 80–6184–69), 2 cm (width) × 32 cm (length) × 1.5 mm (thick) spacers (Catalogue No. 80–6185–64), 20-well × 1.5 mm combs (Catalogue No. 80–6162–08), and some extra 16 cm clamps (Catalogue No. 80–6173–29).

As an alternative to the Hoefer apparatus we also employ Gibco-BRL's Model S2 sequencing gel electrophoresis apparatus (Catalogue No. 21105–036), using spacers either 0.8 or 1.6 mm thick (Catalogue Nos 31109–010 and 31109–028, respectively) and the corresponding 32-well analytical Delrin® combs (Catalogue Nos 21035–118 for 0.8 mm thick; and 21035–126 for 1.6 mm thick). We have also used Gibco BRL's Model SA-60 adjustable sequencing gel electrophoresis system (Catalogue No. 31096–043) using spacers 1.6 mm thick (Catalogue No. 21093–125) and an 18-well analytical Delrin® comb (Catalogue No. 11092–111). Both the Model S2 and SA systems produce longer gels (40 cm and 60 cm, respectively) than the Hoefer SE 620 system (~30 cm) and so allow for better resolution whilst still retaining all of the radioactivity on the gels (this is important for quantitation purposes, see step 9 of the postlabelling assay, below). However, whilst the Model S2 system produces a 30 × 40 cm gel that easily fits standard 35 × 43 cm autoradiographic cassettes and film, the Model SA-60 system produces a 17 × 60 cm gel that requires cutting to fit standard autoradiographic equipment, and does not allow for as many samples to be run per gel (~7) compared with the S2 (~16) and SE 620 systems (2 × ~7).

Preparation
1. Assemble the glass plates with the spacers as described by the respective manufacturers. Ensure that the glass plates are clean and have no cracks.
2. The following instructions are for the preparation of 80 ml of polyacrylamide gel solution, which is sufficient for a single Hoefer SE 620 system gel. To prepare solutions for the larger gel volumes of the Gibco BRL S2 and SA systems (210 and 180 ml, respectively, using 1.6 mm spacers) scale up the ingredients accordingly.

Add together 8 ml of 10 × TBE, 40 ml of the 19:1 acrylamide:bisacrylamide solution and 31.5 g of urea (electrophoresis grade). Stir the mixture at room temperature until the urea is completely dissolved (30–45 min). Next, add 1.6 ml of 3% ammonium persulfate solution and sufficient distilled water to make a total volume of 80 ml, and place on ice.

3. (*Optional.*) Filter the whole mixture through a 0.22 micron filter into a clean bottle and hold on ice. (We use a Falcon 7105 bottle top filter from Becton Dickinson Labware.)

4. Add 25 μl of tetramethylethylenediamine (TEMED) to the cooled gel solution, allow it to mix thoroughly, and pour the gel solution at a steady flow rate between the glass plates. Use either a 25 ml pipette fitted with a manual Pipette Pump$^{®}$ (Bel-Art Products) or an empty 50 ml syringe barrel to direct the flow of the solution close to one of the spacers in order to reduce formation of bubbles.

5. Stop pouring the solution when it is about 1–2 cm from the top of the glass plates and insert the comb between the glass plates. (Try to avoid forming bubbles under the teeth of the comb.) For the Gibco BRL Model S2 and SA systems, strong 'bulldog' clips should be placed on the sides of the plates adjacent to the wells and down both sides of the plates, to ensure a good seal between the plates and the spacers, and over the top edge of the plates to ensure a good seal between the plates and the comb. As the gel solution polymerizes the volume shrinks, so continue to add gel solution to maintain the level close to the top of the plates. Whilst the gel can be used within 2–3 h, ideally it should be left overnight to ensure that polymerization is complete. Once solidified, the gel can be kept at room temperature for 2–3 days, sealed with saran wrap or cling film.

6. To remove the comb, soak the area of the comb with 1X TBE, then carefully 'ease' the comb out. Scrape off any excess gel material from above the wells and on top of the glass plates, and assemble the electrophoresis apparatus according to the respective manufacturer's instructions. Pour 1X TBE into the buffer chambers and ensure that there is no leakage. Clean out debris from the wells of the gel using a syringe and needle, taking care not to puncture the bottom of the wells.

7. Connect the power cables and pre-run the gel for 45 min at the appropriate voltage (see below). (A few microlitres of gel loading buffer can be run in the outer wells to ensure that the gel is running properly.)

8. Load the samples into the wells using either a standard air displacement pipette with 200 μl fine pipette tips or a Hamilton syringe. Avoid using

the outer two or three wells on either side of the gel, since samples applied in these wells may migrate to the edge of the gel and be lost.

Postlabelling assay

1. Digest damaged and undamaged DNA samples (typically 10μg, at 1μg/μl in distilled water) to mononucleosides and damage-containing dimer species by an overnight incubation (16–18 h) at 37°C with 0.4 units of DNase I, 0.044 units of snake venom phosphodiesterase, and 0.4 units of shrimp alkaline phosphatase in 30μl of the DNA digestion buffer. (If several (n) samples are to be digested simultaneously, premix enough 10X TE, MgCl$_2$, enzymes and water to digest $n + 1$ samples, and add an aliquot of this concentrate buffer/ enzyme mixture (typically 20 μ) to each sample.

2. After digestion, precipitate the enzymes by addition of three volumes of ice-cold ethanol and storage at −20°C for 1 h. Pellet the precipitate by centrifugation in a microcentrifuge (21 000 g, 15 min, 4°C) and transfer the supernatant to screw-top microcentrifuge tubes.

3. Lyophilize the supernatants to dryness and dissolve the resulting residues in distilled water (100 μl). Heat the samples in screw-top microcentrifuge tubes (tightly sealed) at 100°C for 15 min to inactivate residual nuclease and phosphatase activity. The samples can then be stored at −20°C.

4. To ^{32}P-phosphorylate the damage-containing dimer species in 5μl of the 'neat' (or diluted) digested DNA solution, add 1μl of 10X T4 polynucleotide kinase buffer, 0.5μl (5 μCi, 1.67 pmol) of [γ-^{32}P]ATP, 7.5 units of T4 polynucleotide kinase and distilled water to a total volume of 10 μl, and incubate the samples for 1 h at 37°C. (Again, if several (n) samples are to be phosphorylated simultaneously, premix enough of the 10X kinase buffer, [γ-^{32}P]ATP, enzyme and water to phosphorylate $n + 1$ samples, and then add 5μl of this mixture to each sample.] When handling the polynucleotide kinase, keep it on ice at all times since the enzyme loses activity fairly rapidly.

5. To consume the excess [γ-^{32}P]ATP, add 1 μl of oligo(dT)$_{16}$ (5 AU$_{260}$ units/ml) and 3.75 units of T4 polynucleotide kinase to each sample and incubate at 37°C for a further 30 min. (Again, make up enough of a T4PNK+oligo mix sufficient for $n + 1$ samples, and add an aliquot to each sample.) Alternatively, add 2 μl of apyrase (0.1 units/μl) to each sample and incubate at 37°C for a further 30 min. Do not combine the T4PNK+oligo(dT)-16-mediated removal of [γ-^{32}P]ATP and the apyrase treatment, as the preparations of apyrase contain nuclease activity which results in the appearance of a ladder stemming from partial digestion of the labelled (dT)$_{16}$ oligomer. (Control experiments have shown that apyrase does *not* affect the damage-containing dimer species.)

6. Add 10 μl of gel loading buffer to each sample, and run half of the reaction mixture on a 20% polyacrylamide/7 M urea gel (prepared as described above). The other half of the reaction mixture can be stored at −20°C. For standard conditions in which all the radioactivity is retained on the gel (including the fast migrating inorganic phosphate bands), electrophoresis is carried out at 800 V for the Hoefer SE 620 system and 1100 V for the Gibco BRL S2 system until the bromophenol blue marker has migrated 11–12 cm (2–4 h); longer times are required for the Gibco BRL SA-60 system. Gels can be run longer to increase separation between bands, but inorganic phosphate will run off the gel.

7. After the electrophoresis is complete, remove gel plates, and wrap the gel in saran wrap or clingfilm. Apply two or three fluorescent markers to areas of the wrapped gel where no radioactivity is expected (e.g. corners of the gel) and expose the gel to fluorescent light for 1 min to activate the fluorescent markers. Visualize the radiolabelled products by autoradiography on Kodak X-Omat K film (30 × 24 cm), X-Omat LS film (30.5 × 25.4 cm) or X-Omat AR film (40 × 30 cm). Autoradiography usually takes 45–90 min. Alternatively, the radiolabelled products can be detected and quantified by phosphoimaging.

8. To isolate radiolabelled material from the gel, locate the radioactive products in the gel using the fluorescent markers to position a template over the gel and excise the radioactive

bands from the gel. (The template can be generated by placing a transparent acetate sheet on the autoradiograph, marking the location of the fluorescent markers and cutting holes in the acetate sheet at the sites of the radioactive bands.) The radioactivity in each gel fragment can be counted without addition of scintillant.

9. If all the radioactivity has been retained on the gel, the molar quantity of material in a particular band can be quantified by determining the fraction of the lane's total radioactivity present in that band, and multiplying this by the molar quantity of ATP used in the reaction. However, before this can be translated into the molar quantity of the lesion in the DNA, it is necessary to determine: (i) the resistance of the phosphodiester bond 5′ to the lesion to digestion by snake venom phosphodiesterase (see Discussion); (ii) the labelling efficiency of the 'dinucleoside' mono-

    phosphate species (see Discussion); and (iii) the amount of DNA in the digested mixture (see step 10).

10. The amount of DNA recovered following digestion of the DNA and removal of the protein (steps 2 and 3) can be accurately determined via HPLC, by comparing the nucleoside content of the digested DNA with standards. With a diode array UV detector it is possible to quantify as little as 0.1–1 nmol of each nucleoside.

HPLC analysis of radioactive bands and nearest-neighbour analysis

The bands in the gel often contain more than one radiolabelled product. These can be isolated by eluting the radioactive products from the gel and resolving them by reverse-phase HPLC. The labelled products have the general structure $^{32}pNpX$, i.e. the lesion is attached 3′ to a normal nucleotide, and it is the latter that bears the labelled phosphate group. The normal nucleotide can of course be any one of the four nucleotides found in DNA. To establish the nature of the nucleotide neighbouring the lesion, the molecules isolated by HPLC can be digested by nuclease P1 to release the labelled 5′-mononucleotides, which can then be identified by reverse-phase HPLC.

Eluting radioactive material from polyacrylamide pieces

1. Mash up the individual polyacrylamide fragments in their scintillation vials and then add 0.5–1 ml of water. (Do not add the water first because the gel becomes too slippery to mash.) Shake the gel vigorously overnight at 4°C.

2. Remove the particulate by filtration and/or centrifugation and rinse the debris with a further 0.5 ml of water. Combine the filtrate and count the recovered radioactivity. (Approximately 50% of the radioactivity should be recoverable.)

HPLC analysis

Instrumentation   A standard instrument is required, capable of pumping a solvent gradient, with online UV detector and radioactivity monitor. If an online radioactivity monitor is not available, fractions can be collected and counted. We have used reverse-phase C_{18} columns purchased from a variety of suppliers (Waters, Supelco and Phenomenex) and have found that they perform similarly. It is useful to have injections loops with different loading capacities, including larger volumes such as 1 or 2 ml.

Elution conditions   Prior to injection on to the column, dilute 5000–50 000 cpm of recovered material to 1 ml with water. (The samples eluted from the polyacrylamide gel contain a large quantity of urea: the radioactive samples should be diluted to reduce the disruptive influence of urea on hydrophobic interactions.) Using a Waters μBondapak C_{18} RCM 8 × 10 RadialPak cartridge, the gradient conditions are as follows: 100% buffer A (50 mM NaH_2PO_4, pH 4.5) and 0% buffer B (100 mM NaH_2PO_4, pH 4.5/ methanol, 1:1 v/v) for 1 min, followed by a linear gradient to 20% buffer A/ 80% buffer B over 30 min at a flow rate of 1 ml/min. Fractions can be collected every 15–30 s.

Nuclease P1 digestion

Having collected the desired fractions from the HPLC, raise the pH to 7.0 by careful addition of 1 M NaOH, dry the solution under vacuum and redissolve the residue in 0.5 ml of distilled water. Desalt the sample by passage through a C_{18} Sep-Pak cartridge (Waters), i.e. apply the sample to an activated (2 ml of methanol) and washed (5 ml of distilled water) cartridge, elute the salt with 1.5 ml of distilled water and elute the radioactive material with 2–3 ml of 50% aqueous methanol. (The best way to monitor the elution of radioactive material is with a Geiger counter.) Remove the solvent and redissolve the residue in 50 μl of nuclease P1 buffer. Add 2 μl of an aqueous solution containing a UV-detectable quantity of unlabelled dinucleotide, to serve as a control of successful digestion of the labelled material, and 1 μl (0.6 unit) of nuclease P1. Allow the reaction to proceed at 37°C for 1 h, and then analyse the digestion products by reverse-phase HPLC, using the same system as that described above.

Marker compounds

The preparation of marker compounds for products detectable by this post-labelling methodology has been described in several articles. Below is a list of compounds and the references detailing their synthesis:

- 'Dinucleoside' monophosphate containing a thymine glycol (NpT^g) – Baleja et al. (1993) and Weinfeld and Soderlind (1991).
- Nucleoside 3'-phosphoglycolate (Npg) – Urata and Akagi (1993) and Henner et al. (1983).
- 'Dinucleoside' monophosphate containing an abasic site (NpS) – Weinfeld et al. (1990) and Weinfeld et al. (1989).

DISCUSSION

Notes on the protocols (troubleshooting)

Heat inactivation of proteins
We have found that simple ethanol precipitation does not always entirely remove enzyme activity, especially phosphodiesterase activity, and consequently the heat inactivation step is required. (Failure to inactivate residual phosphodiesterase activity results in the appearance of a ladder stemming from the partial digestion of the labelled oligo(dT)$_{16}$.) However, if the lesions to be detected are heat labile, some alternatives to heat inactivation are the use of micro-ultrafiltration units (<10 kDa molecular weight cutoff), such as those supplied by Millipore or Micron Separations Inc., or the addition of proteinase K to deactivate the proteins and then ethanol precipitation (Bykov et al., 1995).

Failure to label products efficiently
This can result if (i) there is insufficient [γ-^{32}P]ATP due to its decomposition or the presence of high levels of certain damage over consuming the

[γ-^{32}P]ATP (see discussion on quantitative aspects, below), or (ii) the kinase has little or no activity in the reaction mixture. The latter can occur if the protein itself is inactive or if there is an inhibitor of kinase activity. We have observed that it is important to keep the total salt concentration, including Tris-HCl, in the kinase reaction low, preferably < 80 mM.

Spurious bands in untreated DNA

- This can be genuine damage already present in the DNA. We have observed high levels of damage, including abasic sites, in commercial preparations of herring sperm DNA.
- We have also observed that the use of certain commercial preparations of T4PNK increases the likelihood of spurious bands. In a series of experiments studying damaged DNA samples, and control samples containing no DNA, we have found that T4PNK purchased from Promega (coupled with an apyrase-mediated removal of excess [γ-^{32}P]ATP) yields the lowest level of 'background' activity in the control samples whilst still maintaining high levels of damage-labelling. We therefore recommend the use of this particular enzyme (together with the apyrase step) when attempting to measure low levels of oxidative damage. However, one advantage of the T4PNK+oligo-mediated removal of excess [γ-^{32}P]ATP is that it provides a check of the kinase reaction, thus substantiating that a low number of lesions detected is in fact due to a low level of damage, and not due to poor kinase activity or decomposed [γ-^{32}P]ATP.
- Old preparations of phenol used in the purification of cellular DNA or divalent metal ion contamination of DNA may give rise to additional/adventitious oxidative damage.

Quantitative aspects of the assay

Quantitative parameters arising from the dependence of postlabelling assays on several enzymes has been critically reviewed by Hemminki *et al.* (1993). In our procedure the key enzymes are snake venom phosphodiesterase and polynucleotide kinase. With snake venom phosphodiesterase, the question arises as to whether the phosphodiester bond immediately 5' to the lesion is totally refractory to the enzyme. In studies with 5'-^{32}P end-labelled thymine glycol-containing 'dinucleotide' model compounds, we have shown that there is an approximate 50% loss of the dimer species using the conditions described above, and that this matches a relative loss of thymine glycol-containing dimer species observed after overnight digestion of DNA samples, compared with identical samples digested for shorter time periods (Routledge *et al.*, 1998). This indicates that the phosphodiester bond 5′ to the glycol-bearing nucleoside is not totally resistant to digestion, and therefore we can assume that the SVPD-postlabelling assay (including an overnight digestion) underestimates the level of thymine glycol. In contrast, the phosphoglycolate-containing species appear far more resistant to overnight digestion. However, as pointed out by Hemminki *et al.* (1993), results obtained with dinucleotides may not necessarily reflect results obtained with longer oligonucleotides and DNA.

In the case of the T4PNK-mediated labelling reaction, studies with model compounds have shown the labelling efficiencies of the thymine glycol-containing dimer species (dNpT^g) to be similar to that of the equivalent umodified dimers (dNpT) (Routledge *et al.*, 1998), suggesting that the damage-containing dimer species are good substrates for T4PNK and efficiently labelled. In contrast, thymine glycol-containing nucleoside 3′-monophosphates (T^{g}p) (the corresponding products of MN+CSPD digestion in the Randerath postlabelling protocol) have been shown to be poor substrates for T4PNK, poorer than normal thymidine 3′-monophosphates (Tp), and consequently have very low labelling efficiencies (Hegi *et al.*, 1989). The superior labelling of the damage-containing dimer species (dNpT^g) is because it is a normal nucleoside 3′-monophosphate moiety that is labelled, whilst with the damaged nucleoside 3′-monophosphate species (T^{g}p) the kinase has to act directly on the modified nucleoside. Based upon these model studies, the damage-containing dimer species (dNpX) derived from SVPD-postlabelling digestion are expected to be efficiently labelled.

In the labelling reaction it is important to have at least a three- to five-fold excess of [γ-^{32}P]ATP in the reaction mixture. Therfore, with samples containing high levels of damage it is sometimes necessary to dilute the digested DNA solutions to maintain an excess of [γ-^{32}P]ATP in the labelling reaction. Low levels of [γ-^{32}P]ATP can result in preferential labelling of certain substrates. For example, we have observed that the phosphoglycolate-containing species appear to be better substrates for T4PNK than the thymine glycol-containing species.

As mentioned above, we have found that using T4PNK (from Promega) coupled with an apyrase-mediated removal of excess [γ-^{32}P]ATP yields the lowest levels of background activity for control samples containing no DNA, markedly improving the detection limits. Using this procedure, we have detected basal levels of thymine glycol and phosphoglycolate lesions in human peripheral blood lymphocyte DNA via SVPD-postlabelling (≤ 20 fmol/μg DNA assayed for thymine glycols and ≤ 22 fmol/μg DNA assayed for phosphoglycolates) (Routledge *et al.*, 1998). However, given the possible underestimation of lesions (particularly thymine glycols, see above) these values should be taken as lower limits.

Clearly, control experiments using standard oligonucleotides or DNAs containing a known number of lesions must be conducted to determine correction factors required for the valid quantitation of lesions detected by SVPD-postlabelling. These studies are currently underway but, for now, the technique must be considered *semi*-quantitative.

General comments on the assay (advantages and disadvantages)

Certain advantages of our approach to postlabelling are mentioned in the Introduction, including an absence of labelling of undamaged DNA, improved labelling efficiency of lesion-containing molecules, and nullification of adventitious oxidative effects during labelling. There are, however, several drawbacks.

1. In common with other postlabelling techniques, a significant problem is that unless marker compounds are available, the assay provides no structural information to help to identify the detectable lesions. (However, a good number of the marker compounds for the oxidative lesions detected have been prepared: see marker compounds section, above).
2. Since the damaged base or sugar moiety is isolated with a normal neighbouring 5′-nucleotide attached, there are always four species generated for each lesion. (However, this does allow for the determination of the 5′-nearest neighbour; see section on nearest-neighbour analysis, above).
3. Only a limited number of oxidative lesions can be detected by this technique, with 8-oxoguanine and 5-hydroxymethyluracil being amongst the common lesions not detectable by our approach. (However, given that the assay yields four products per lesion, the detection of only a limited number of lesions dramatically simplifies what would otherwise be a very complex situation). Studies are presently underway to improve the detection capabilities of the assay.

Despite the above mentioned disadvantages, the SVPD-postlabelling is an important addition to the methodologies used to assess oxidative DNA damage. In particular it has several advantages over existing postlabelling methods, including its capacity to detect lesions that cannot be measured by the other approaches or that are detected with low efficiency.

ACKNOWLEDGEMENTS

We wish to acknowledge the people who have contributed to the development of this assay, particularly Dr. Michel Liuzzi, Krista-June Soderlind, Dr Malcolm Paterson, Dr Garry Buchko, Jane Lee, Dr Michael Routledge, Neil Bennett, Dr. Karen Bowman and Lynda Dickinson. Work described here was supported by the Ministry of Agriculture, Fisheries and Food (UK), the National Cancer Institute of Canada, and the Medical Research Council (UK), with further support from the Canadian Cancer Society and the Alberta Cancer Board.

REFERENCES

Baleja, J.D., Buchko, G.W., Weinfeld, M. and Sykes, B.D. (1993) Characterization of γ-radiation induced decomposition products of thymidine-containing dinucleoside monophosphates by NMR spectroscopy. *J. Biomol. Struct. Dynam.*, **10**, 747–762.

Bertoncini, C.R.A. and Meneghini, R. (1995) DNA strand breaks produced by oxidative stress in mammalian cells exhibit 3′-phosphoglycolate termini. *Nucleic Acids Res.*, **23**, 2995–3002.

Buchko, G.W. and Weinfeld, M. (1993), Influence of nitrogen, oxygen, and nitroimidazole radiosensitizers on DNA damage induced by ionizing radiation. *Biochemistry*, **32**, 2186–2193.

Bykov, V.J., Kumar, R., Forsti, A. and Hemminki, K. (1995) Analysis of UV-induced DNA photoproducts by P-32 postlabeling. *Carcinogenesis*, **16**, 113–118.

Hegi, M.E., Sagelsdorff, P. and Lutz, W. (1989) Detection by ^{32}P-postlabeling of thymidine glycol in γ-irradiated DNA. *Carcinogenesis*, **10**, 43–47.

Hemminki, K., Försti, A., Löfgren, M., Segerbäck, D., Vaca, C. and Vodicka, P. (1993) Testing of quantitative parameters in the ^{32}P-postlabelling method. In: Phillips, D.H., Castegnaro, M. and Bartsch, H. (eds) *Postlabelling Methods for the Detection of DNA Adduct*, IARC Scientific Publications No. 124, pp. 51–63. IARC Publications, Lyons.

Henner, W.D., Rodriguez, L.O., Hecht, S.M. and Haseltine, W.A. (1983), γ Ray induced deoxyribonucleic acid strand breaks. *J. Biol. Chem.*, **258**, 711–713.

Jones, G.D.D. and Weinfeld, M. (1996) Dual action of Tirapazamine in the induction of DNA strand breaks. *Cancer Research*, **56**, 1584–1590.

Jones, G.D.D., Dickinson, L., Lunec, J. and Routledge, M.N. (1999), SVPD postlabelling detection of oxidative damage negates the problem of adventitious damage formation during ^{32}P-labeling. *Carcinogenesis*, **20**, 503–507.

Liuzzi, M., Weinfeld, M. and Paterson, M.C. (1989) Enzymatic analysis of isomeric trithymidylates containing ultra violet light-induced cyclobutane pyrimidine dimers I. Nuclease P1-mediated hydrolysis of the intradimer phosphodiester linkage. *J. Biol. Chem.*, **264**, 6355–6363.

Möller, L. and Hofer, T. (1997), [^{32}P]ATP mediates formation of 8-hydroxy-2′-deoxyguanosine from 2′-deoxyguanosine, a possible problem in the ^{32}P-postlabelling assay. *Carcinogenesis*, **18**, 2415–2419.

Möller, L., Zeisig, M. and Vodicka, P. (1993) Optimization of an HPLC method for analysis of ^{32}P-postlabeled DNA adducts. *Carcinogenesis*, **14**, 1343–1348.

Phillips, D.H. (1997), Detection of DNA modifications by the ^{32}P-postlabelling assay. *Mutation Res.*, **378**, 1–12.

Podmore, K., Farmer, P.B., Herbert, K.E., Jones, G.D.D. and Martin, E.A. (1997), ^{32}P-postlabelling approaches for the detection of 8-oxo-2'-deoxyguanosine-3'-monophosphate in DNA. *Mutation Res.*, **178**, 139–149.

Povey, A.C., Wilson, V.L., Weston, A., Doan, V.T., Wood, M.L., Essigmann, J.M. and Shields, P.G. (1993), Detection of oxidative damage by ^{32}P-postlabelling: 8-hydroxydeoxyguanosine as a marker of exposure. In: Phillips, D.H., Castegnaro, M. and Bartsch, H. (eds) *Postlabelling Methods for the Detection of DNA Adducts*, IARC Scientific Publications No. 124, pp. 105–114. IARC Publications, Lyons.

Randerath, K., Reddy, M.V. and Gupta, R.C. (1981) ^{32}P-labeling test for DNA damage. *Proc. Natl. Acad. Sci. USA*, **78**, 6126–6129.

Reddy, M.V., Bleicher, W.T. and Blackburn, G.R. (1991) ^{32}P-Postlabeling detection of thymine glycols: evaluation of adduct recoveries after enhancement with affinity chromatography, nuclease P1, nuclease S1, and polynucleotide kinase. *Cancer Commns*, **3**, 109–117.

Routledge, M.N., Lunec, J., Bennett, N. and Jones G.D.D. (1998) Measurement of basal levels of oxidative damage in human lymphocyte DNA by ^{32}P-postlabelling, *Proc. Am. Assn Cancer Res.*, **39**, 287.

Schuler, D., Ottender, M., Sagelsdorff, P., Edar, E., Gupta, R.C. and Lutz, W.K. (1997) Comparative analysis of 8-oxo-2'-deoxyguanosine in DNA by ^{32}P- and ^{33}P-postlabelling and electrochemical detection. *Carcinogenesis*, **18**, 2367–2371.

Stuart, G.R. and Chambers, R.W. (1987), Synthesis and properties of oligodeoxyribonucleotides with an AP site at a preselected position. *Nucleic Acids Res.*, **15**, 7451–7462.

Urata, H. and Akagi, M. (1993) A convenient synthesis of oligonucleotides with a 3'-phosphoglycolate and 3'-phosphoglycoaldehyde terminus. *Tetrahedron Lett.*, **34**, 4015–4018.

Weinfeld, M. and Buchko, G.W. (1993) Postlabelling methods for the detection of apurinic sites and radiation-induced DNA damage. In: Phillips, D.H., Castegnaro, M. and Bartsch, H. (eds) *Postlabelling Methods for the Detection of DNA Adducts*, IARC Scientific Publications No. 124, pp. 95–103. IARC Publications, Lyons.

Weinfeld, M. and Soderlind, K.-J.M. (1991) ^{32}P-Postlabeling detection of radiation-induced DNA damage: identification and estimation of thymine glycols and phosphoglycolate termini. *Biochemistry*, **30**, 1091–1097.

Weinfeld, M., Liuzzi, M. and Paterson, M.C. (1989), Selective hydrolysis by exo- and endonucleases of phosphodiester bonds adjacent to an apurinic site. *Nucleic Acids Res.*, **17**, 3735–3745.

Weinfeld, M., Liuzzi, M. and Paterson, M.C. (1990) Response of phage T4 polynucleotide kinase toward dinucleotides containing apurinic sites: design of a ^{32}P-postlabeling assay for apurinic sites in DNA. *Biochemistry*, **29**, 1737–1743.

Weinfeld, M., Soderlind, K.-J.M. and Buchko, G.W. (1993), Influence of nucleic acid base aromaticity on substrate reactivity with enzymes acting on single-stranded DNA. *Nucleic Acids Res.*, **21**, 621–626.

Weinfeld, M., Luizzi, M. and Jones, G.D.D. (1996), A postlabeling assay for oxidative DNA damage. In: Pfeifer, G.P. (ed.) *Technologies for Detection of DNA Damage and Mutations*, p. 63. Plenum Press, New York.

10 Mapping Reactive Oxygen-induced DNA Damage at Nucleotide Resolution

Henry Rodriguez and Steven A. Akman

OXIDATIVE DNA DAMAGE

Exposure of DNA to fluxes of reactive oxygen species (ROS) induces a variety of DNA base modifications, strand breaks and DNA-protein cross-links (reviewed in Halliwell and Aruoma, 1991). ROS-induced DNA damage may be an important intermediate in the pathogenesis of human conditions such as cancer and ageing (Ames, 1987; Guyton and Kensler, 1993). ROS-induced DNA damage products are both mutagenic and cytotoxic (reviewed in Wallace, 1994).

The mutational spectra of ROS, e.g. H_2O_2 (Moraes *et al.*, 1989; Akman *et al.*, 1991) and the transition metal ions Fe, Cu (Loeb *et al.*, 1988; Tkeshelasnvili *et al.*, 1991), and Cr (Kawanishi *et al.*, 1994) have been studied in model systems, but the relationship of induced DNA damage to these spectra remains unknown. Knowledge of this relationship is crucial in order to extrapolate from model mutational analysis systems to endogenous mutational spectra generated *in vivo*. This is of significant scientific and medical interest, since elucidation of the role that oxidatively induced DNA damage has in carcinogenesis could ultimately enable the development of rational therapeutic interventions that might be beneficial in the prevention of human malignancies. Progress in this area has been hampered until recently by the lack of damage measurement techniques with nucleotide resolution.

NUCLEOTIDE RESOLUTION MAPPING OF ROS-INDUCED DNA DAMAGE BY THE LIGATION-MEDIATED POLYMERASE CHAIN REACTION

The ligation-mediated polymerase chain reaction (LMPCR) is a genomic sequencing method for mapping of rare DNA single-stranded breaks

Measuring in vivo *Oxidative Damage: A Practical Approach.* Edited by J. Lunec and H. R. Griffiths. © 2000 by John Wiley & Sons, Ltd. ISBN 0 471 81848 8.

(Mueller and Wold, 1989; Pfeifer *et al.*, 1991). Briefly, LMPCR is a six-step process (Rodriguez *et al.*, 1995), as shown in Figure 10. 1. The steps are as follows:

1. conversion of a modified base (base damage) into a strand break, either chemically or enzymatically, followed by primer extension of an annealed gene-specific oligonucleotide (upstream primer 1) to generate blunt ends;
2. ligation of a universal asymmetric double-strand linker on to the blunt ends;

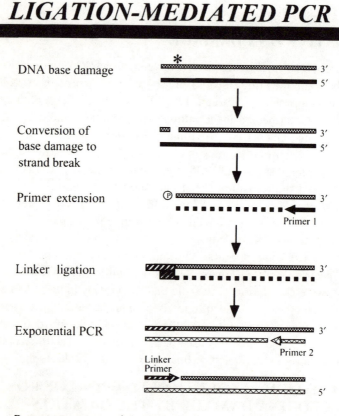

Run on sequence gel

Transfer onto Nytran membrane

Hybridize with nested single-stranded labelled probe

Figure 10.1   Schematic representation of the steps in DNA base damage mapping by LMPCR.

3. PCR amplification using a second gene-specific oligonucleotide (upstream primer 2) along with a linker primer (downstream linker primer);
4. separation of the DNA fragments on a sequencing polyacrylamide gel;
5. transfer of the DNA to a nylon membrane by electroblotting;
6. hybridization of a radiolabelled probe prepared by repeated primer extension using a third gene-specific oligonucleotide (upstream primer 3).

LMPCR is a versatile technique, in that induced DNA strand breaks can be mapped as long as strand break induction results in a $5'$-phosphoryl end. Therefore, LMPCR can be used in sensitively mapping DNA base modifications, e.g. cyclopyrimidine dimers (Pfeifer *et al.*, 1991), alkylated bases (Lee *et al.*, 1994), or adducted bases (Denissenko *et al.*, 1996), which can be converted to strand breaks with $5'$-phosphoryl ends by chemical or enzymatic means.

The LMPCR technique has been described in detail elsewhere (Rodriguez *et al.*, 1995). Key steps with recent modifications (such as the incorporation of a hot-start by using AmpliTaq® Gold polymerase, providing better signal-to-noise ratio) are as follows.

Primer extension

1. Primer 1 is extended in a siliconized 0.65 ml tube: a thermocycler (MJ Research Inc., Watertown, Massachusetts is used for all incubations.
2. DNA (0.5–1.3 μg) is diluted in a volume of 15 μl of a solution containing 40 mM Tris-HCl, pH 7.7, 50 mM NaCl and 0.75 pmol of Primer 1.
3. DNA is denatured at 98°C for 3 min and the primer (Primer 1) annealed at 45°C for 30 min (primers for the *PGK1* housekeeping gene are in Rodriguez *et al.*, 1995, and *p53* gene in Tornaletti *et al.*, 1993).
4. After cooling the sample to 4°C, 9 μl of the following mix is added: 7.5 μl MgCl$_2$-dNTP mix (20 mM MgCl$_2$, 20 mM dithiothreitol and 0.25 mM of each dNTP), 1.1 μl dH$_2$O and

0.4 μl of Sequenase® 2.0 (13 units/μl, US Biochemicals, Cleveland, Ohio).
5. Samples are then incubated at 48°C for 15 min.
6. The samples are placed on ice and 6 μl ice-cold 310 mM Tris-HCl, pH 7.7, is added.
7. To inactivate Sequenase®, samples are incubated at 67°C for 15 min.
8. Samples are placed on ice.

Ligation

1. The primer-extended molecules that have a 5′ phosphate are ligated to an unphosphorylated synthetic asymmetric double-stranded linker (Gao et al., 1994).
2. To each sample (consisting of 30 μl), 45 μl of the following ligation mix is added: 13.33 mM MgCl$_2$, 30 mM dithiothreitol, 1.7 mM ATP, 83.3 μg/ml BSA, 100 pmol linker and 5 units of T$_4$ DNA ligase (5 units/μl, Boehringer Mannheim, Gaithersburg, Maryland).
3. Samples are incubated overnight at 18°C.
4. Ligase is inactivated by incubation at 70°C for 10 min.
5. Samples are placed on ice.
6. Next, add 25 μl of 10 M ammonium acetate, 1 μl of 0.5 M EDTA, pH 8.0, 1 μl of 20 μg/μl glycogen, followed by 250 μl of ice-cold ethanol to precipitate the DNA.
7. DNA pellets are redissolved in 50 μl dH$_2$O.

PCR amplification

1. To each sample, 50 μl of an AmpliTaq® Gold polymerase mix (2X AmpliTaq® Gold reaction buffer (Perkin Elmer Inc., Foster City, California), 1 mM MgCl2, 400 μM of each dNTP, 10 pmol primer 2, 10 pmol of linker primer (Gao et al., 1994) and 3.0 units of AmpliTaq® Gold polymerase (5 units/μl, Perkin Elmer Inc., Foster City, California) is added, and reactions are overlaid with mineral oil. (Note: 2X AmpliTaq® Gold buffer contains 3 mM MgCl$_2$; therefore, the final MgCl$_2$ concentration in the 50 μl 2X AmpliTaq® Gold polymerase mix is 4 mM, due to the added 1 mM MgCl$_2$.

This translates into a 2 mM final MgCl2 concentration in the 100 μl PCR reaction).

2. Reactions undergo one PCR cycle of 95°C for 15 min (activating AmpliTaq® Gold polymerase), T_m of Primer 2 for 2 min, and 72°C for 3 min; 18 PCR cycles of 95°C for 1 min, 1°C below T_m of Primer 2 for 2 min, and 72°C for 3 min. Lastly, an extension is performed at 72°C for 10 min.

3. Following the PCR reaction, a stop mix (13 μl of 3 M sodium acetate, pH 5.2, 3 μl of 0.5 M EDTA, pH 8.0, and 9 μl of dH$_2$O) is added under the mineral oil layer.

4. Samples are extracted with 170 μl of premixed phenol:chloroform (50 μl:120 μl), then ethanol is precipitated by adding 370 μl ice-cold ethanol.

5. Air-dried DNA pellets are dissolved in 7.0 μl of premixed formamide-dye (2.3 μl dH$_2$O, 4.7 μl formamide loading dye: 95% v/v formamide, 10 mM EDTA, pH 8, 0.05% w/v xylene cyanol, 0.05% w/v bromophenol blue) in preparation for sequencing gel electrophoresis.

We have adapted LMPCR to map a broad spectrum of ROS-induced DNA base modifications – those that are sensitive to the DNA glycosylase and abasic lyase activities of the Nth protein (also known as endonuclease III) and Fpg protein (also known as formamidopyrimidine DNA glycosylase) of *Escherichia coli* (Rodriguez *et al.*, 1995; a detailed experimental protocol is given in this reference). Using LMPCR, we have studied DNA damage caused by transition metal ion-catalysed reduction of H$_2$O$_2$ in the presence of ascorbate (Rodriguez *et al.*, 1995, 1997a,b).

The distribution of oxidative damage induced in exons 5 and 9 of human *p53*, as well as the promoter region of human *PGK1*, was assessed. The autoradiograms indicating the damage distributions induced in the region of the human *PGK1* gene covered by primer set A (see Rodriguez *et al.*, 1995, for primer set designations) and in exon 9 of the human *p53* gene are shown in Figure 10.2. In these regions of the genome, the base damage frequency distributions induced in dialysed DNA *in vitro* by Cu(II) and Fe(III) plus ascorbate/H$_2$O$_2$ are nearly identical. Sucrose was included in the Fe(III) reaction to suppress the direct strand break signal. Sucrose had no effect on the LMPCR-derived damage distribution signals indicated by

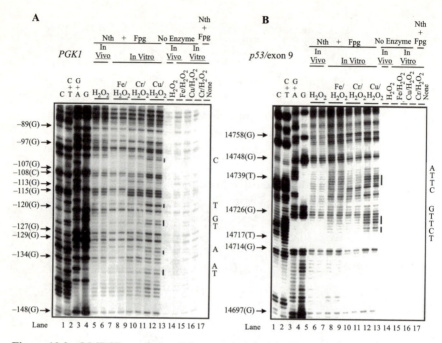

Figure 10.2 LMPCR analysis of damage induced in: (A) the promoter region of human *PGK1* using primer set A (transcribed strand); and (B) exon 9 of human *p53* (transcribed strand). *Lanes 1–4*: DNA treated with standard Maxam–Gilbert cleavage reactions. *Lanes 5–6, 13*: DNA recovered from intact human foreskin fibroblasts exposed to 50 mM H_2O_2. *Lanes 7–8, 14*: dialysed genomic DNA treated with 100 μM Fe(III)/100 μM ascorbate/5 mM H_2O_2 in the presence of 0.3 M sucrose. *Lanes 9–10, 15*: DNA treated with 50 μM Cu(II)/100 μM ascorbate/5 mM H_2O_2 in the presence of 1 mM potassium phosphate buffer. *Lanes 11–12, 16*: DNA treated with 50 μM Cr(VI)/100 μM ascorbate/5 mM H_2O_2. *Lane 17*: DNA incubated in potassium phosphate buffer and digested with Nth and Fpg proteins. The DNA in lanes 5–12 was digested with Nth and Fpg proteins after treatment; the DNA in lanes 13–16 was incubated in digestion buffer alone after treatment. Positions of high damage frequency bases are marked with arrows to the left of lane 1. The sequence of positions heavily damaged in the presence of chromium, but not copper or iron, is denoted by rectangles to the right of lane 12. Reproduced with permission.

comparison of *in vitro* Cu(II)/ascorbate/H_2O_2 reactions conducted in either 1 mM potassium phosphate buffer or 0.3 M sucrose (Rodriguez *et al.*, 1997b). The base damage distribution associated with these two transition metal ions is non-uniform. Prominent base damage hotspots are observed in *PGK1* and in *p53* exon 9 (Figure 10.2).

The distribution of damage caused by Cr(VI)/ascorbate/H_2O_2 was similar, but not identical, to that mediated by copper or iron ions in these regions. The unique chromium ion-sensitive positions were often thymines, whereas thymines were uncommonly modified in the presence of Fe(III) or Cu(II) (Figure 10.2).

The distribution of DNA base damage occurring *in vivo* induced by exposure of cultured human male fibroblasts to 50 mM H_2O_2 was also determined in *PGK 1* and *p53* (Rodriguez *et al.*, 1997b). Here, 50 mM H_2O_2 induces a global damage frequency in human male fibroblast DNA *in vivo* equivalent to the damage frequency induced *in vitro* by 50 μM Cu(II)/ 100 μM ascorbate/5 mM H_2O_2 (data not shown). Figure 10.2 demonstrates that the base damage distribution induced *in vivo* in the assessed regions of *PGK1* and *p53* was identical to the damage distribution induced *in vitro* by Cu(II) or Fe(III) plus H_2O_2/ascorbate and similar to the damage distribution induced *in vitro* by Cr(VI)/H_2O_2/ascorbate. Control experiments, such as extending the number of cell rinses with Chelex®-treated phosphate-buffered saline after H_2O_2 exposure to as many as nine times, resulted in no alteration of the damage frequency. This demonstrates that the putative *in vivo* damage distribution was not a post-DNA extraction artefact. Furthermore, the concentration of H_2O_2 in the medium at the end of the 30 min incubation period was well below that required to observe damage signals by LMPCR.

Comparison of damage intensity among sequence contexts was made by applying sequential Wilcoxson rank-sum tests (Rodriguez *et al.*, 1997b). Guanine is the most easily modified base associated with H_2O_2-mediated DNA damaging reactions both *in vivo* and *in vitro*, with the triplet d(CGC) being the principal hotspot sequence.

Exposure of human male fibroblasts to a concentration of H_2O_2 several orders of magnitude higher than those generated under basal metabolic conditions was required in order to generate sufficient DNA base damage for the purpose of damage frequency mapping by LMPCR. Therefore, we assessed to what extent artefacts caused by exposure to high-concentration H_2O_2 contributed non-physiological distortion of the observed damage frequency patterns. The principal effect of high-concentration H_2O_2, causing severe oxidative stress, turned out to be release of cellular transition metal ions from normally sequestered extranuclear sites.

Evidence for release of transition metal ions from extranuclear sites by the severe oxidative stress was obtained by assessing DNA damage in isolated human male fibroblast nuclei. DNA damage was assessed globally by neutral denaturing agarose gel electrophoresis (Drouin *et al.*, 1996). Figure 10.3

shows that isolated nuclei behave similar to naked DNA in that neither H_2O_2 alone nor Cu(II)/Fe(III) + ascorbate causes detectable DNA damage in isolated human male fibroblast nuclei. However, nuclei behave differently from naked DNA in that significant DNA damage was observed if H_2O_2 and Cu(II)/Fe(III) were added. In fact, base damage induced by H_2O_2/copper ion in the isolated nuclei was greater than that induced *in vivo* by equimolar

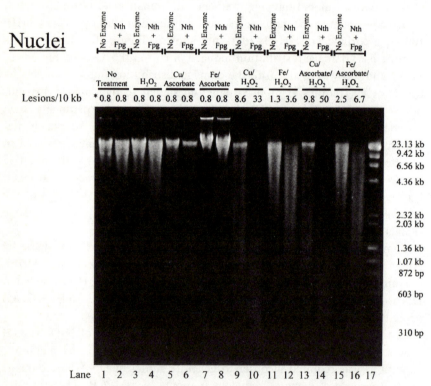

Figure 10.3 Global frequency of direct strand breaks (*'No Enzyme' lanes*) and direct strand breaks plus modified bases (*'Nth + Fpg' lanes*) observed after exposure of isolated human male fibroblast nuclei to 50 mM H_2O_2 (lanes 3–4), 50 μM Cu(II) + 100 μM ascorbate (lanes 5–6), 50 μM Fe(III) + 100 μM ascorbate (lanes 7–8), or combinations of these reagents (lanes 9–16). *Lane 17*: 500 ng lambda DNA digested with *Hin*d III and 500 ng PhiX174 digested with *Hae* III. *Lanes 1–2*, nuclei not treated (Controls). Nuclear isolation was performed as described in Rodriguez *et al.*, 1997b. Exposures were for 30 min at 37°C, after which DNA was isolated. Electrophoresis was carried out as described in Drouin *et al.*, 1996. (*) 0.8 is a lower limit of estimated lesion frequency in these lanes; a more precise value could not be determined.

H_2O_2 with or without supplemental $50\,\mu M$ Cu(II) in the medium (data not shown). Thus, isolated nuclei have the following two notable properties:

1. They contain endogenous reducing agents capable of reducing transition metals such that the metals redox cycle in the presence of H_2O_2.
2. They do not contain sufficient bound metals (or metal-like ligands) to cause significant base damage in the presence of H_2O_2.

ENHANCEMENT OF LMPCR DAMAGE DETECTION SENSITIVITY BY GENOMIC GENE ENRICHMENT

The requirement for exposure of fibroblasts to cytotoxic concentrations of H_2O_2 in order to map ROS-induced DNA base damage *in vivo* by LMPCR spurred us to develop methods to enhance the sensitivity of LMPCR. One approach to enhancing LMPCR-generated base damage signal intensity is to increase the relative copy number of the target gene in the substrate genomic DNA. This was accomplished by size-fractionating restriction endonuclease-digested, ROS-exposed genomic DNA by continuous elution electrophoresis (CEE) through a preparative agarose gel (Rodriguez and Akman, 1998) (Figure 10.4) as follows.

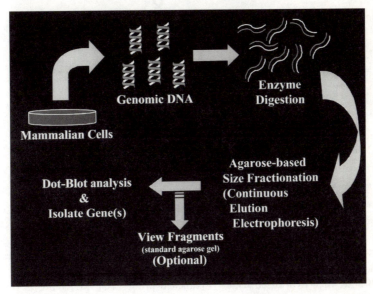

Figure 10.4   Schematic representation of the steps in target gene enrichment by CEE.

1. Total human genomic DNA was damaged by exposure to $0.5\,\text{mM}$ H_2O_2 in the presence of $50\,\mu\text{M}$ Cu(II) and $100\,\mu\text{M}$ ascorbate, then digested with *Bam*H I.
2. Aliquots of $300\,\mu\text{g}$ were fractionated by CEE.
3. Fractions of 1 ml were collected, viewed on agarose gels (an optional step – Figure 10.5A,B), and fractions containing the *p53* (exons 3–11) and *PGK1* genes were determined by dot blot analysis (Figure 10.5C,5D) and then used as LMPCR substrates.

This procedure yielded ~$12\,\mu\text{g}$ of *p53* or *PGK1*–enriched DNA, respectively; the representation of each gene was enhanced ~$25 - fold$.

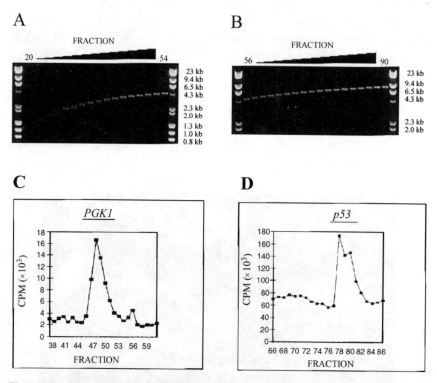

Figure 10.5 Determination of CEE fractions containing target genes: $300\,\mu\text{g}$ aliquots of *Bam*H I-digested total human genomic DNA were fractionated by CEE: 90 ng aliquots of every other CEE fraction were analysed by 0.7% agarose gel electrophoresis – gel (A), fractions 20–54; gel (B), fractions 56–90 – to assess the DNA fragment size range per fragment; 30 ng aliquots were dot-blotted to nylon membranes, then hybridized with [^{32}P]-labelled probes: (C): *PGK1* probe; (D): *p53* probe. Hybridization signal intensities were quantified by PhosphorImager analysis.

The CEE procedure has been described in detail (Rodriguez and Akman, 1998). Key steps are as follows.

Bam H I digestion of total genomic DNA

1. Following ROS treatment, 50 μg aliquots of DNA were digested in 250 units *Bam*H I (Gibco BRL, Gaithersburg, Meryland) and 10 μg RNase A (Sigma, St Louis, Missouri), in a volume of 500 μl at 37°C for 4 h. Digestion was carried out in Gibco BRL 1X React 3 enzyme buffer.
2. Digestions were stopped by the addition of 10 μl 0.5 M EDTA.
3. Phenol (AquaPhenol; Appligene/ONCOR, Gaithersburg, Maryland)/phenol-chloroform/chloroform (J.T. Baker, Phillipsburg, New Jersey) extractions were carried out to remove the proteins.
4. DNA was precipitated by the addition of 50 μl of 3 M NaOAc, pH 7, and 1 ml of ice-cold ethanol.
5. Air-dried pellets were resuspended in 50 μl 1X TE (10 mM Tris-HCl, pH 8, 1 mM EDTA), and prepared for continuous elution electrophoresis.

Loading the DNA sample and running the continuous elution electrophoresis apparatus

1. 30 μl of 10X agarose loading dye (Sigma, St Louis, Missouri) was added to 300 μg *Bam*H I-digested (Gibco BRL, Gaithersburg, Maryland) genomic DNA in a volume of 300 μl TE (final concentration of 1 μg/ml).
2. The DNA sample was loaded on to a Model 491 Prep Cell (Bio-Rad Laboratories, Hercules, California), containing a 0.5% preparative agarose gel (SeaKem® Gold; FMC BioProducts, Rockland, Marie) and a 0.25% agarose stacking gel (SeaKem® Gold; FMC BioProducts, Rockland, Maire).
3. The sample was run in 50 mM TBE (Tris-borate/EDTA), at 55 constant volts (PowerPac 300 Power Supply; Bio-Rad Laboratories, Hercules, California) at 4°C.

4. An elution flow rate of 50 μl/min was maintained by a peristaltic pump (Model EP-1 Econo Pump; Bio-Rad Laboratories, Hercules, California), with fraction collection times of 20 min (1 ml final volume).

Fraction viewing by standard agarose electrophoresis (optional)

After continuous elution electrophoresis, 30 μl from each fraction (representative of 90 ng of DNA in 50 mM TBE buffer) was mixed with 3 μl of 10X agarose loading dye (Sigma, St Louis, Missouri) and loaded on to 0.7% standard agarose gels (SeaKem® LE, FMC BioProducts, Rockland, Maire) and run in 50 mM TBE, at 100 constant volts. Gels were stained by containing ethidium bromide (0.5 μg/ml), then photographed using Polaroid 667 film (Polaroid Corporation, Cambridge, Massachusetts). DNA concentration was determined by A_{260} measurements.

Fraction screening for gene of interest by dot-blot analysis

1. After continuous elution electrophoresis, 10 μl from each fraction (representative of 30 ng DNA in 50 mM TBE buffer) was added to 190 μl of 0.4 M NaOH, 10 mM EDTA solution.
2. Samples were heated to 95°C for 5 min in a thermocycler (PTC-100, MJ Research, Watertown, Masachusetts), placed on ice, and loaded on to a Dot-Blot apparatus (Bio-Dot; Bio-Rad Laboratories, Hercules, California) containing a positively charged nylon plus membrane (Qiabrane; Qiagen, Santa Clarita, California).
3. Wells were rinsed with 200 μl of 0.4 M NaOH, 10 mM EDTA solution.
4. Membrane was then soaked in 2X SSC (NaCl/sodium citrate) for 5 min, UV cross-linked (1200 J/m^2) (Stratalinker; Stratagene, La Jolla, California), and placed in a hybridization tube containing hybridization solution (0.25 M NaPO$_4$, 1 mM EDTA, 7% SDS (sodium dodecyl sulfate), 1% BSA (bovine serum albumin) (fraction V; Sigma, St Louis, Missouri)), and radiolabelled probe.
5. This was placed into a hybridization oven (HB 1100D Red Roller II; Pharmacia Biotech, Piscataway, New Jersey) overnight (Rodriguez et al., 1995).

6. After overnight hybridization at 66°C, membranes were washed for 5 min in Buffer A (20 mM NaPO$_4$, 1 mM EDTA, 2.5% SDS, 0.25% BSA (fraction V; Sigma, St Louis, Missouri)) at 66°C, followed by 5 min in Buffer B (20 mM NaPO$_4$, 1 mM EDTA, 1% SDS) at 66°C.

7. Buffer B wash was repeated two times.

8. Air-dried membranes were exposed to Kodak XAR-5 X-ray films (Eastman Kodak, Rochester, New York) with intensifying screens (Optex, Cedar Knolls, New Jersey) at −70°C.

9. The intensity of each dot was quantitated by PhosphorImager analysis (Model 425S; Molecular Dynamics, Sunnyvale, California).

Cleavage of enriched samples at ROS-induced modified bases

1. Following ROS treatment, continuous elution electrophoresis and fraction screening by dot-blot analysis, fractions containing the highest percentage of the gene of interest were precipitated by making two tubes, each consisting of 500 μl eluted fraction, 50 μl of 3M NaOAc, pH 7, 1 μl glycogen (20 μg/μl), and 1 ml ice-cold ethanol.

2. DNA was ethanol precipitated by a 10 min incubation on dry ice.

3. Air-dried pellets were resuspended in 25 μl 1X TE, pH 8. Each respective pair was pooled to yield a 50 μl total volume consisting of 3 μg DNA.

4. Respective fractions (containing 3 μg DNA) were then mixed with 50 μl of a unique 2X Nth/Fpg reaction buffer (91.4 mM Tris-HCl, pH 7.7, 200 mM KCl, 1.1 mM EDTA, 0.2 mM DTT, 200 μg/μl BSA-fraction V), yielding a final volume of 100 μl.

5. To this, 400 ng Fpg protein (formamidopyrimidine DNA glycosylase) from *Escherichia coli* and 100 ng Nth protein (endonuclease III) from *E. coli* were added and samples were digested at 37°C for 60 min as previously described (Rodriguez *et al.*, 1995). Fpg and Nth proteins were provided by Dr Timothy R. O'Connor of the Beckman Research Institute of the City of Hope (Duarte, California).

6. Control samples (no enzyme) were incubated in buffer alone (data not shown).

7. Digestions were terminated as previously described (Rodriguez et al., 1995).

8. Finally, the DNA pellets were dissolved in Sequenase buffer (40 mM Tris-HCl, pH 7.7 and 50 mM NaCl) for LMPCR, as previously described (Rodriguez et al., 1995).

Cleavage of non-enriched samples at ROS-induced modified bases

This procedure has been described in detail elsewhere (Rodriguez et al., 1995). Briefly, following ROS treatment, 10 μg aliquots of DNA were digested in 400 ng Fpg protein (formamidopyrimidine DNA glycosylase) from *Escherichia coli* and 100 ng Nth protein (endonuclease III) from *E. coli*, in a volume of 100 μl at 37°C for 60 min. Digestions were terminated and finally the DNA pellets were dissolved in Sequenase buffer (40 mM Tris-HCl, pH 7.7, and 50 mM NaCl) for LMPCR.

LMPCR analysis

Air-dried membranes were exposed to Kodak XAR-5 X-ray films (Eastman Kodak, Rochester, New York) with intensifying screens (Optex, Cedar Knolls, New Jersey, USA) at −70°C. Each band represents a nucleotide position where damage was converted into a strand break. The signal intensity of each band reflects the number of DNA molecules with ligatable ends terminating at that position; intensity of each band was quantitated by PhosphorImager analysis (Model 425S; Molecular Dynamics, Sunnyvale, Califirnia).

Use of the target gene-enriched DNA as LMPCR substrate resulted in an average 24-fold enhancement of LMPCR-derived base damage signal intensity as compared to non-enriched total genomic DNA (Figure 10.6). The enhancement of LMPCR-derived damage signal intensity by target gene enrichment is sufficient to permit mapping of base damage induced by non-cytotoxic *in vivo* exposures of breast epithelial cells to H_2O_2. In fact, enhancement of signal intensity is such that damage signals may be detected in DNA extracted from cells that have not been exposed to exogenous

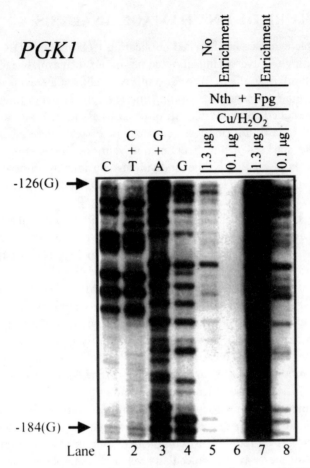

Figure 10.6 Autoradiogram of DNA fragments produced by LMPCR analysis (primer set G, transcribed strand) of damage induced in the promoter region of human *PGK1* by treating purified genomic DNA with 50 μM Cu(II)/100 μM ascorbate/0.5 mM H_2O_2 in phosphate buffer for 30 min at 37°C. *Lanes 5–6*: non-enriched treated DNA, digested with Nth and Fpg proteins after treatment; *lanes 7–8*, treated DNA, digested with Nth and Fpg proteins after treatment, then enriched by CEE fractionation; *lanes 1–4*, non-enriched, non-treated DNA subjected to standard Maxam–Gilbert cleavage reactions.

oxidative stress. It is unclear yet whether these background signals are attributable to endogenous oxidative DNA base damage or are a result of post-DNA extraction artefacts; studies designed to differentiate these possibilities are in progress.

THE FUTURE OF DNA DAMAGE ANALYSIS

Considerable evidence has accrued implicating ROS-induced DNA damage in contributing to adverse clinical outcomes in our increasingly aged population. Concomitantly, the need for and utility of clinical assessment of ROS-induced cellular damage is becoming increasingly evident. Interventional strategies designed to lower the risk of diseases associated with ageng, (e.g. cancer) will require informative measurement of ROS-induced damage as intermediate markers of the effects of interventions. A screening technique needs to be implemented so that individuals can take appropriate steps to prevent significantly high levels of ROS-induced DNA damage from possibly forming.

To date, clinical measurements of cellular oxidative damage have been principally analytical measurements. By-products of lipid peroxidation have been assessed in blood, urine or breath samples by high performance liquid chromatography with electrochemical detection (HPLC–EC) or gas chromatography with mass spectrometric detection (GCMS). Tests for pro-oxidant or antioxidant analytes (e.g. trace metals and vitamins) or antioxidant enzymes (such as superoxide dismutase, catalase and glutathione peroxidase) are commercially available. Finally, *in vivo* oxidative DNA damage has been assessed analytically by measuring urinary excretion of the ROS-induced DNA base modification 8-hydroxydeoxyguanosine (Loft *et al.*, 1993).

These analytical measurements may have an important function as clinical risk assessment and intervention tools; however, they do not provide information about damage at nucleotide resolution. Such fine-structure information will be required to complement our increasing knowledge of disease pathogenesis at the molecular level. For example, in many human cancers we understand that the occurrence of somatic mutations in certain critical genes plays an important pathogenetic role. In order to understand individual risk for somatic mutagenesis in these target genes, it will be imperative to assess promutagenic damage at specific sites in target genes.

Continued development of LMPCR as a sensitive nucleotide-resolution damage assessment tool holds out the possibility that LMPCR will become an important tool for individual damage/risk assessment. Analytical measurements of basal steady-state levels of endogenous ROS-induced DNA damage in human cellular DNA suggest that a target detection sensitivity for LMPCR of 1 lesion per 10^6 DNA bases will be required to bring it on line as a clinically useful tool. At present, LMPCR combined with genomic gene enrichment by CEE can detect 2.0 ROS-induced lesions per 10^5 DNA bases (Rodriguez and Akman, 1998); however, a large amount of DNA is required for the enrichment procedure. Further improvements in sensitivity, e.g. by enriching genomic DNA specifically for damage-containing molecules, may overcome the remaining hurdles and allow LMPCR to progress from a laboratory research tool to a clinical risk assessment tool.

REFERENCES

Akman, S.A., Forrest, G.P., Doroshow, J.H. and Dizdaroglu, M. (1991) Mutation of potassium permanganate- and hydrogen peroxide-treated plasmid pZ189 replicating in CV-1 monkey kidney cells. *Mutat. Res.*, **261**, 123–130.

Ames, B.N., (1987) Oxidative DNA damage, cancer, and aging. *Ann. Intern. Med.*, **107**, 526–545

Calderaro, M., Martins, E.A. and Meneghini, R. (1993) Oxidative stress by menadione affects cellular copper and iron homeostasis. *Mol. Cell Biochem*, **126**, 17–23.

Denissenko, M.F., Pao, A., Tang, M. and Pfeifer, G.P. (1996) Preferential formation of benzo(a)pyrene adducts at lung cancer mutational hotspots in p53. *Science*, **274**, 430–432.

Drouin, R., Rodriguez, H., Gao, S., Gebreyes, Z., O'Connor, T.R., Holmquist, G.P. and Akman, S.A. (1996) Cupric ion/ascorbate/hydrogen peroxide-induced DNA damage: DNA-bound copper ion primarily induces base modifications. *Free Radical Biol. Med.*, **21**, 261–273.

Gao, S., Drouin, R. and Holmquist, G.P. (1994) DNA repair rates mapped along the human PGK1 gene at nucleotide resolution. *Science*, **263**, 1438–1440.

Guyton, K.Z. and Kensler, T.W. (1993) Oxidative mechanisms in carcinogenesis. *Br. Med. Bull.*, **49**, 523–544.

Halliwell, B. and Aruoma, O.I. (1991) DNA damage by oxygen-derived species. *FEBS Lett.*, **281**, 9–19.

Kawanishi, S., Inoue, S. and Yamamoto, K. (1994) Active oxygen species in DNA damage induced by carcinogenic metal compounds. *Environ. Health Perspect.* **102**, Suppl 3, 17–20.

Lee, C.S., Pfeifer, G.P. and Gibson, N.W. (1994) Mapping of DNA alkylation sites induced by aziridinylbenzoquinones in human cells by ligation-mediated polymerase chain reaction. *Cancer Res.*, **54**, 1622–1626.

Loeb, L.A., James, E.A., Waltersdorph, A.M. and Klebanoff, S.J. (1988) Mutagenesis by the auto-oxidation of iron with isolated DNA. *Proc. Natl Acad. Sci. USA*, **85**, 3918–3922.

Loft, S., Fischer-Nelsen, A., Jeding, I.B., Vistisen, K. and Poulsen, H.E. (1993) 8-hydroxydeoxyguanosine as a urinary biomarker of oxidative DNA damage. *J. Toxicol. Environ. Health*, **40**, 391–404.

Moraes, E.C., Keyse, S.M., Pidoux, M. and Tyrrell, R.M. (1989) The spectrum of mutations generated by passage of a hydrogen peroxide damaged shuttle vector plasmid through a mammalian host. *Nucleic Acids Res.*, **17**, 8301–8312.

Mueller, P.R. and Wold, B. (1989) *In vivo* footprinting of a muscle-specific enhancer by ligation-mediated PCR. *Science*, **246**, 780–786.

Pfeifer, G.P., Steigerwald, S.D., Mueller, P.R., Wold, B. and Riggs, A.D. (1989) Genomic sequencing and methylation analysis by ligation-mediated PCR. *Science*, **246**, 810–813.

Pfeifer, G.P., Drouin, R., Riggs, A.D. and Holmquist, G.P. (1991) *In vivo* mapping of a DNA adduct at nucleotide resolution: detection of pyrimidine (6–4) pyrimidone photoproducts by ligation-mediated polymerase chain reaction. *Proc. Natl Acad. Sci. USA*, **88**, 1374–1378.

Rodriguez, H. and Akman, S.A. (1998) Large scale isolation of genes as DNA fragment lengths by continuous elution electrophoresis through an agarose matrix. *Electrophoresis*, **19**, 646–652.

Rodriguez, H., Drouin, R., Holmquist, G.P., O'Connor, T.R., Boiteux, S., Laval, J., Doroshow, J.H. and Akman, S.A. (1995) Mapping of copper/hydrogen peroxide-induced DNA damage at nucleotide resolution in human genomic DNA by ligation-mediated polymerase chain reaction. *J. Biol. Chem.*, **270**, 17633–17640.

Rodriguez, H., Drouin, R., Holmquist, G.P. and Akman, S.A. (1997a) A hot spot for hydrogen peroxide-induced damage in the human hypoxia-inducible factor 1 binding site of the PGK1 gene. *Arch. Biochem. Biophys*, **338**, 207–212.

Rodriguez, H., Holmquist, G.P., D'Agostino, R. Jr, Keller, J. and Akman, S.A. (1997b) Metal ion-dependent hydrogen peroxide-induced DNA damage is more sequence specific than metal specific. *Cancer Res.*, **57**, 2394–2401.

Tornaletti, S., Rozek, D. and Pfeifer, G.P. (1993) The distribution of UV photoproducts along the human *p53* gene and its relation to mutations in skin cancer. *Oncogene*, **8**, 2051–2057.

Tkeshelasnvili, L.K., McBride, T., Spence, K. and Loeb, L.A. (1991) Mutation spectrum of copper-induced DNA damage. *J. Biol. Chem.*, **266**, 6401–6406.

Wallace, S.S. (1994) DNA damage processed by base excision repair: biological consequences. *Int. J. Radiat. Biol.*, **66**, 579–584.

Part V   Antioxidant Activity

11 Measurement of Plasma Antioxidant Activity

S.R.J. Maxwell

INTRODUCTION

Recent years have seen an increasing acceptance that the pathophysiology of a variety of human diseases involves an excess of oxidative damage to molecules and cells in biological systems (Figure 11.1). An imbalance of oxidative stresses and antioxidant defence has been implicated in diseases such as atherosclerosis, reperfusion injury, diabetes mellitus, inflammation and cancers. Patients who are deficient in antioxidants show evidence of increased oxidative damage *in vivo*, and low antioxidant status has now been linked to the development of diseases in epidemiological studies. These observations have stimulated interest in the possibility that measurement of antioxidant status may have a role in the clinical arena. Measurements of antioxidant function might help to detect individuals and biological systems where antioxidant defences are compromised and the risk of oxidative damage is highest. This is particularly important because of the availability of natural antioxidant supplements such as vitamin C, vitamin E, ubiquinol and selenium as well as antioxidant drugs that might improve antioxidant function *in vivo*.

PRINCIPLES OF MEASURING 'TOTAL' ANTIOXIDANT ACTIVITY

A measurement of the capacity of a biological system to withstand oxidative attack (its antioxidant activity) might offer important information about its health and viability. Furthermore, since some radical-scavenging (chain-breaking) antioxidant molecules are consumed in the process of protecting against oxidants, changes in antioxidant activity may also reflect previous or ongoing oxidative stress.

Measuring in vivo *Oxidative Damage: A Practical Approach*. Edited by J. Lunec and H. R. Griffiths. © 2000 by John Wiley & Sons, Ltd. ISBN 0 471 81848 8.

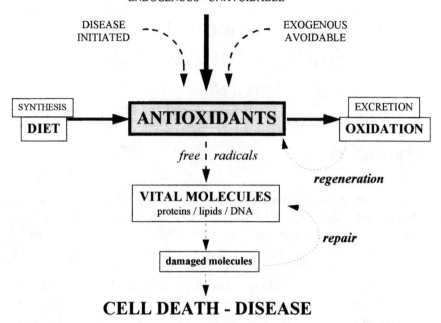

Figure 11.1 The importance of pro-oxidant–antioxidant balance. Much of the oxidative stress faced by cells is an unavoidable result of their own aerobic metabolism. Oxidative stress may also be initiated by certain pathophysiological states such as inflammation, ischaemia-reperfusion, hyperglycaemia and iron overload. Furthermore, a variety of exogenous sources of avoidable oxidative stress exist, including ionizing radiation and cigarette smoke. In health, most of the stresses will be absorbed by the endogenous antioxidant defences. A small proportion of the stress will result in damage to vital molecules. However, as long as the rate of damage is limited, these molecules can be repaired or replaced. When the rate of damage exceeds this capability, then cell death occurs. The level of antioxidant defence is dynamic and influenced by several factors, including the rate of dietary intake of antioxidants and their precursors, the rate of endogenous synthesis of antioxidants, the rate at which they are being oxidized and the rate of excretion. The interpretation of instantaneous antioxidant measurements may depend on all of these factors.

The task of quantifying total antioxidant activity could be approached in two different ways. Firstly, the concentrations of all of the individual molecules that are currently recognized as antioxidants could be measured. This would present several problems:

- It would be time consuming, expensive and technically demanding.
- It might fail to yield information about their combined effective-ness, i.e. the 'total' antioxidant activity may be greater than the sum of the individual antioxidants because of cooperative interactions.
- It might not account for the influence of antioxidant substances that are as yet undiscovered or are technically difficult to assay.

A second, simpler approach to the problem of quantifying antioxidant activity in test samples is to subject them to controlled oxidative stress and measure either the rate and extent of oxidation or how long it takes for oxidation to begin. This can then be standardized against a known antioxidant. This concept is attractive because it avoids many of the problems above and, since the assays are dynamic and involve the application of an oxidative stress, they might also be viewed as 'physiological', i.e. they measure the functional aspects of antioxidant activity.

Two important variables that will influence the result of this approach are the kind of *oxidative stress* that is imposed and the *index of oxidative damage* used to define the course of radical-induced oxidation (Figure 11.2). Firstly, if a radical generating system based on free transition metal ions is chosen as the source of oxidative stress, then the greatest antioxidant activity will be found in samples with the greatest capacity to sequestrate these ions and prevent radical generation. Conversely, if oxidative stress is based on the spontaneous decomposition of aqueous radical generating compounds, then test samples well endowed with aqueous radical-scavenging antioxidant molecules will appear to have high antioxidant activity. Secondly, the index of oxidative damage used to define free radical-induced oxidation may be an important influence on the outcome of the assay. If damage to aqueous proteins is the marker for oxidative damage, then it is less likely that radical-scavenging antioxidant molecules in the lipid phase will have any influence on the capacity to prevent it. Alternatively if the marker of damage is the pro-pagation of lipid peroxidation chains, then these same molecules might be the most influential. Accordingly, any assay to measure 'total' antioxidant activity will be limited by the oxidant stress it imposes and the index of damage that is chosen, and the results must be interpreted in that light (see below).

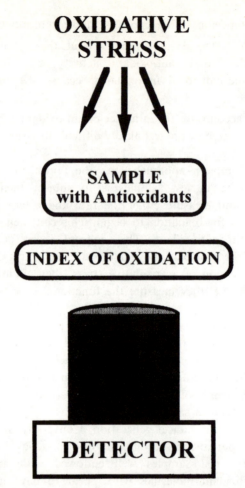

Figure 11.2   General principles of antioxidant assays. Oxidative stress is applied to a test sample containing antioxidants. The index of oxidation (e.g. lipid peroxidation) is then detected in the test sample and its rate compared with that in a known standard antioxidant sample.

TOTAL ANTIOXIDANT ASSAYS

This chapter will concentrate on methods of measuring free radical-scavenging activity in plasma and other aqueous biological fluids. A variety of methods employing different sources of oxidative stress and detection have now been described (Table 11.1). The initial assays relied on tissue homogenization and subsequent spontaneous autoxidation. The first widely employed technique was the TRAP assay, which was based on the spontaneous

Table 11.1 Methods of measuring free radical scavenging antioxidant activity

Oxidative stress	Oxidation index	Measurement	Reference
Tissue autoxidation (brain)	Lipid peroxidation	TBARS	Stocks et al., 1974
Peroxyl radicals (ABAP)	Lipid peroxidation	Oxygen consumption	Wayner et al., 1987
Peroxyl radicals (ABAP)	Luminol	Luminescence	Metsä-Ketelä, 1991
Peroxyl radicals (AMVN)	Luminol	Luminescence	Metsä-Ketelä and Kirkkola, 1992
H_2O_2/peroxidase	o-Phenylenediamine	Absorbance 430 nm	Nakamura et al., 1987
Peroxyl radicals (AAPH)	β-Phycoerythrin	Fluorescence 565 nm	DeLange, 1989
Cu^{2+}/H_2O_2	Lipid peroxidation	TBARS	Arshad et al., 1991
Cu^{2+}/hydroperoxide	cis-Parinaric acid	Fluorescence 413 nm	McKenna et al., 1991
H_2O_2/peroxidase/enhancer	Luminol	Luminescence	Whitehead et al., 1992
Peroxyl radicals (AAPH)	β-Phycoerythrin	Fluorescence 565 nm	Cao et al., 1993
Ferryl myoglobin	ABTS	Absorbance 734 nm	Miller et al., 1993
Fe^{3+}	Fe/tripyridyltriazine	Absorbance at 593 nm	Benzie and Strain, 1996

ABTS = 2,2-azinobis-(3-ethylbenzothiazoline-6-sulphonate)
AAPH = 2,2′-azobis(2-amidinopropane) dihydrochloride
ABAP = 2,2′-azobis(2-amidopropane) hydrochloride
AMVN = 2,2′-azobis(2,4-dimethylvaleronitrile)
TBARS = thiobarbituric acid reactive substances
DOPA = 3,4-dihydroxyphenylalanine
AMVN = 2,2′-azobis(2,4-dimethylvaleronitrile).

decomposition of the radical-generating compounds as a source of oxidative stress. However, the assay was time-consuming and the oxygen electrode detector was rather unstable over that time period. Subsequent modifications of the TRAP assay employed more reliable and precise chemiluminescent detection. The expanding interest in the application of antioxidant assays led to the development of several newer methods. These assays have various advantages, including improved reproducibility and sensitivity, reduced measurement times, lack of a requirement for close temperature control of the assay reagents and the potential for automation.

INFLUENCES UPON PLASMA 'TOTAL' ANTIOXIDANT ACTIVITY

Any measurement of plasma antioxidant status may be influenced by several factors (Table 11.2). The intake of radical-scavenging antioxidants in the diet (e.g. vitamin C, vitamin E, flavonoids) has received much attention. However, some important endogenous antioxidants are 'waste-products' of endogenous metabolism (e.g. urate and bilirubin) and others result from de novo synthesis (e.g. plasma protein thiols). Both processes may be indirectly influenced by the diet. The situation is further complicated by the fact that some antioxidants are also continuously regenerated (e.g. vitamin C, vitamin E). In some

Table 11.2 Factors influencing antioxidant status

Factor	Comment
Genetic	Genetic defects in coding for antioxidant enzymes
Diet	Radical-scavenging molecules (e.g. vitamin C, vitamin E, flavonoids)
	Trace metal cofactors of antioxidant enzymes (e.g. selenium)
Malabsorption	Fat-soluble vitamin E uptake reduced in pancreatic disease (e.g.)
Transport	Vitamin E not transported in abeta-lipoproteinaemia (e.g.)
Rate of consumption	Radical scavenging antioxidants are consumed by oxidative stress
Rate of regeneration	Some antioxidants are regenerated *in vivo* (e.g. vitamin C, vitamin E)
Rate of elimination	Some antioxidants are excreted (e.g. urate, vitamin C)
Inactivation	Antioxidant enzymes may be inactivated by oxidative stress
Adaptation	In some circumstances antioxidant defences may be upregulated in response to persistent oxidative stress

malabsorbative states, such as pancreatic disease, the uptake of essential dietary constituents may be impaired in spite of adequate dietary provision.

The steady-state levels of radical-scavenging antioxidants are determined not only by their rate of intake but also by their rate of consumption by continuous free radical oxidative stress or their rate of elimination (Figure 11.1). Most notably, urate and vitamin C are excreted in the kidney and bilirubin is excreted in the bile. Failure of either system may have a dramatic impact on plasma total antioxidant activity via non-redox mechanisms. Since the enzyme and preventive antioxidants are also important in defending against oxidative stress, defects in either may have an impact on radical-scavenging antioxidant status. Genetic factors and dietary intake of trace elements that are essential cofactors (e.g. selenium, copper, zinc, manganese) may influence the activity of antioxidant enzymes. The complexity of natural antioxidant defence has been extensively reviewed elsewhere (Halliwell, 1990; Halliwell and Gutteridge, 1990; Stocker and Frei, 1991).

INTERPRETATION OF 'TOTAL' ANTIOXIDANT ASSAYS

In addition to all of the above influences, there remain further uncertainties about the interpretation of measurements of 'total' antioxidant activity in plasma. A central question is whether it is possible to make a meaningful interpretation of antioxidant status from plasma samples. This may be a relatively crude measure of antioxidant protection at sites relevant to pathophysiological processes. These are often microenvironments that are probably not in equilibrium with the bulk of the extracellular fluids.

Another major disadvantage of measuring 'total' antioxidant activity is that the quantification is highly dependent on the assay used. Each has its own specific oxidative stress and index marker of oxidation. As a result of these differences each seems to be more readily influenced by one antioxidant component than the others (Table 11.3). It is unknown at present which individual antioxidants are *qualitatively* the most important *in vivo* but it is likely to differ depending on the stresses imposed. Therefore, it is also unclear which antioxidant assay offers the most relevant prognostic information. This issue may only be settled if prospective studies establish

Table 11.3 Comparison of the TRAP, ECL and ABTS antioxidant assays (data for the antioxidant assays has been derived from Wayner *et al.*, 1987; Maxwell, 1995; Goode *et al.*, 1995; Miller *et al.*, 1993)

	TRAP	ECL	ABTS
Normal range (μM)[a]	$410_{\pm}74$	$424_{\pm}86$	$1460_{\pm}70$
Inter-individual variability (CV) (%)	18.0	20.3	4.8
Reproducibility (CV)			
Within-batch (%)	3.4	1.7	1.6
Between batch (%)	8.3	5.0	6.1
Stoichiometry (*n*-values)[a]			
Urate	0.65	1.0	1.02
Ascorbate	0.85	1.0	0.99
Vitamin E	1.0	0.76–1.2	0.97
Bilirubin	–	0.74–1.0	1.50
Free thiols	0.16	0.33–0.56	0.90
Protein thiols	0.16	0	0.63[c]
Contributions (%)			
Urate	58	70	33
Ascorbate	14	10	9
Vitamin E	7	9	3
Proteins[b]	21	0	43
Others	0	11	12

[a] Values for total antioxidant activity and stoichiometric (*n*-values) of individual antioxidants are described in relation to trolox ('trolox equivalence') which has a stoichiometric value of 1.0.
[b] Percentage contribution of proteins for the TRAP assay includes 'plasma proteins' (presumably exclusive of vitamin E), and for the ABTS assay includes 'albumin'.
[c] Value for albumin in solution.

that one of the available total antioxidant measurements has a superior correlation with a clinical outcome. Indeed, it should be acknowledged that the oxidative stress used in most antioxidant assays is not relevant to normal human physiology where the relevant oxidants are $O_2^{\bullet-}$, H_2O_2, $OH^{\bullet}$, HOCl and ferryl haem proteins.

For these reasons it is unlikely that measurements of plasma total antioxidant activity will make an impact in the clinical arena in the forseeable future, although they may represent a valuable research tool. In particular, those assays that are sensitive and reproducible may be well suited to particular situations such as observing relatively small acute changes in radical-scavenging activity in pathophysiological situations (e.g. exercise, reperfusion injury, acute inflammation) and measuring the impact of drugs or nutritional supplements with putative antioxidant activity on radical-scavenging status *in vivo*.

THE TRAP ASSAY

The total peroxyl radical-trapping antioxidant parameter (TRAP) assay was originally described by Wayner *et al.* (1987) and has probably been the most widely used assay for radical-scavenging antioxidant status. The method is based on the constant production of water-soluble peroxyl radicals by the spontaneous temperature-dependent decomposition of an 'initiator' such as 2,2′-azobis(2-amidopropane) hydrochloride (ABAP). The peroxyl radicals generated can abstract hydrogen atoms from organic compounds (either plasma lipids themselves or linoleic acid added to the test sample) to initiate lipid peroxidation. If the test sample added to the system contains antioxidants that can scavenge peroxyl radicals, then the onset of lipid peroxidation will be delayed until the antioxidants have been consumed (Figure 11.3). Lipid peroxidation is sensed by the increased rate of oxygen consumption as indicated by the fall in oxygen saturation measured by an oxygen electrode. Given a constant rate of peroxyl radical production, the delay will be proportional to the peroxyl scavenging capacity of the test solution.

Reagents

- 2,2′-azobis-2-amidinopropane hydrochloride (ABAP) (Polysciences, Warrington, Pensylvania). Prepare a 0.4 M ABAP stock solution by dissolving in 10.0 ml phosphate buffered saline fresh each day.
- Linoleic acid (Sigma, Poole, Dorset).
- Phosphate buffered saline (PBS, 10 mM phosphate buffer, pH 7.4) (Sigma, Poole, Dorset).

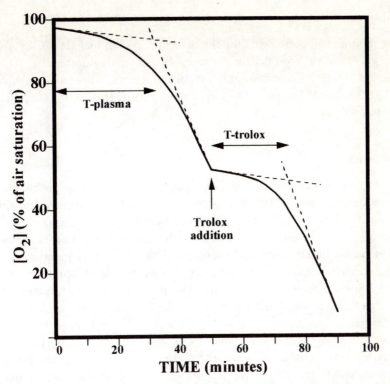

Figure 11.3   The TRAP assay. At 37°C ABAP (2,2'-azobis(2-amidopropane) hydro-chloride) spontaneously decomposes to produce peroxyl radicals at a known rate. At time 0 the reaction is initiated in the presence of a volume of plasma supplemented with linoleic acid. Lipid peroxidation in this sample is indicated by the uptake of oxygen as sensed by an oxygen electrode. The period during which peroxidation is resisted as a result of chain-breaking antioxidant activity in the sample (*t*-plasma) is most adequately indicted by the tangential lines to the oxygen saturation curve. This is then compared with a similar retardation of oxygen uptake produced by a standard antioxidant solution of Trolox (*t*-trolox). Adapted from Wayner *et al.* (1987).

Calibration standards

Trolox (6-hydroxy-2,5,7,8-tetramethylchroman-2-carboxylic acid) (Sigma, Poole, Dorset), is a water-soluble vitamin E analogue that is relatively stable in aqueous solution and forms a convenient and widely used standard for antioxidant assays. A 0.4 mM solution is prepared by dissolving 25 mg Trolox (mol. wt. 250.29) in 1 ml of ethanol and making up to 250 ml in PBS. This stock solution can be divided into 2 ml aliquots and stored frozen. The Trolox standard is stable for at least 6 months at $-20°$.

Apparatus

The minimum requirements for the detection system include a Clarke-type oxygen electrode mounted in a lucite plunger, a thermostated ($\pm 0.05°C$) reaction vessel, an amplifier and a recording device. These are available in a YSI Biological Oxygen Monitor (Yellow Spings Instrument Co., Yellow Springs, Ohio).

Procedure

The oxygen electrode should be inspected for the presence of air bubbles or the accumulation of surface oxides. The temperature controlling device should be set to 37°C. PBS (3.0 ml) is added to the oxygen electrode cell and allowed to equilibrate to the bath temperature. The electrode is inserted and baseline oxygen is monitored to ensure that there is no consumption in the absence of the initiator. The oxygen electrode is calibrated to read 100% at atmospheric pressure. The plasma samples are prepared by vortex mixing a 100 μl aliquot of plasma and 4μl of linoleic acid in a small vial for 30 s.

Add 30 μl of ABAP stock solution to the cell. A 50 μl portion of the prepared plasma mixture is added to the 3.0 ml of PBS now containing 4.0 mM ABAP at 37°C in the oxygen electrode cell. The rate of oxygen uptake is monitored until it reaches its maximum after approximately 30 min. The system is then calibrated by the addition of 25 μl of the Trolox standard solution (0.4 mM). The rate of oxygen consumption is again monitored until it has reached a second peak (Figure 11.3). It is important to protect the contents of the cell from room light during the experiment since the initiator is subject to photodecomposition.

Calculations

The induction time (t) or delay before achieving the maximum rate of oxygen uptake after the addition of plasma or Trolox can be calculated by drawing the tangent to the oxygen saturation curve immediately after sample addition and after it has reached a maximum. The time from sample addition to the point of intersection is 't' (Figure 11.3). The delay in lipid peroxidation produced by the test plasma (t-plasma) is compared with the delay associated with the Trolox standard (t-trolox). The TRAP result (μ mol peroxyl radicals scavenged per litre of plasma) is calculated from the equation:

$$AOA_{TRAP} = \frac{2.0\,[\text{trolox}] \times t\text{-plasma} \times V\text{-trolox}}{t\text{-trolox} \times V\text{-plasma}}$$

where [trolox] is the concentration of the Trolox standard (0.4 mM) and V is the sample volume. A factor of 2.0 is included in the equation because Trolox can scavenge two peroxyl radicals per molecule according to the reactions:

$$ROO^{\bullet} + TH \longrightarrow ROOH + T^{\bullet}$$
$$ROO^{\bullet} + T^{\bullet} \longrightarrow ROOT$$
$$\text{Net} \quad 2ROO^{\bullet} + TH \longrightarrow ROOH + ROOT$$

where $ROO^{\bullet}$ is a peroxyl radical and TH is Trolox in its normal reduced form.

Reference range

The mean TRAP value ($\pm$ SD) for 45 human plasma samples was 820 $\pm148\mu$M (Wayner et al., 1987). It has been calculated that the percentage contribution (mean$\pm$ SD) of the individual antioxidants to the total TRAP is: urate $58 \pm 18\%$, thiol groups $21 \pm 10\%$, ascorbate $14 \pm 8\%$ and vitamin E $7 \pm 2\%$. The within-batch and between-batch variability have been estimated to be 3.4% and 8.3% (CV) respectively.

Interference

EDTA-plasma samples give values that are 10% less than corresponding heparinized plasma samples.

Comments

The TRAP assay has some major disadvantages. The assays are time-consuming (taking up to 90 min to measure a single sample), which precludes the analysis of large sample numbers. The assay also requires a considerable degree of expertise and training. Furthermore, the oxygen electrode sensor may not be able to maintain its stability over such a long time period, introducing a high degree of analytical imprecision. The apparent reliance upon urate as a major influence on the total antioxidant activity of plasma may be perceived as a problem.

Some of these problems have been addressed by more recent modifications of the original assay (Metsä-Ketelä, 1991; Metsä-Ketelä and Kirkkola, 1992). This involved changing the detection system from an oxygen electrode to luminol chemiluminescence, which offers more precision. The peroxyl radicals produced by ABAP oxidize luminol when all antioxidants have been consumed, leading to the generation of light – the duration of the delay to light emission being proportional to the radical-trapping activity of the sample antioxidants. A further modification of the assay was made to enable the measurement of antioxidants in the lipid phase. This involved changing to the lipid-soluble initiator 2,2′-azobis(2,4-dimethylvaleronitrile) (AMVN) as the source of peroxyl radicals. Using a chemiluminescent end-point, it is possible to study antioxidant activity in lipoproteins and cell membranes.

THE ENHANCED CHEMILUMINESCENT (ECL) ASSAY

The enhanced chemiluminescent antioxidant assay was originally described by Whitehead *et al.* (1992). Chemiluminescence may be defined as the emission of photons of light by electrons in an excited state and may follow the oxidation of a variety of compounds. Such reactions can be catalysed by the enzyme horseradish peroxidase (HRP), which behaves as an oxidoreductase with a wide variety of possible substrates but is relatively specific for H_2O_2 as the hydrogen acceptor. During the catalytic cycle of the enzyme, H_2O_2 first oxidizes the enzyme to give a derivative known as Compound I, which retains the oxidizing equivalents of the hydroperoxide. The two electrons are then replaced in two one-electron steps, with the first creating an intermediate between HRP and Compound I, known as Compound II, and the second regenerating the native enzyme. For each of these electron transfers the chemiluminescent substrate (e.g. luminol) can act as the electron donor and is converted to its radical form. These radicals then undergo a series of further reactions, producing intermediates in electronically excited states that decay to emit photons of light, which can be detected by a conventional luminometer.

Unfortunately, the oxidation of luminol by Compound II to regenerate HRP is rather slow and the light emissions are often of relatively low intensity and decay rapidly. In enhanced chemilu-minescence (ECL), substituted phenolic compounds such as para-iodophenol known as *enhancers* greatly increase the intensity and duration of light emission: they accelerate the conversion of Compound II back to the active HRP enzyme by offering a more favourable substrate for oxidation. The enhancer radical can then itself oxidize luminol while being reduced back to its non-radical form. Thus the enhancer is not consumed in the reaction but regenerated for further oxidation catalysed by HRP.

The generation of light by ECL depends on the continuous production of free radical intermediates of para-iodophenol. Therefore, the light emission is sensitive to interference by rad-ical-scavenging antioxidants but is restored when any added anti-oxidants have been consumed in the reaction. If the generation of radical intermediates is constant, then the time period of light suppression is directly related to the amount of radical-scaven-ging antioxidants added into the reaction mixture.

Reagents

1. Enhanced chemiluminescence reagents are obtained as 'Amerlite Signal Reagent', (B.M. Browne UK Ltd, Reading). These consist of Amerlite signal reagent buffer (pH 8.5, 30 ml) and tablets A and B (containing luminol, para-iodophenol and sodium perborate). Sodium perborate dissolves in aqu-eous solution to yield H_2O_2, the primary oxidant substrate for HRP. Amerlite signal reagent is prepared by adding tablets A and B to the signal reagent buffer solution and allowing them to dissolve. The signal reagent can be divided into aliquots and frozen at $-20°C$ for long periods prior to use.

2. Horseradish peroxidase conjugate (anti-mouse immunoglo-bulin HRP-linked whole antibody from sheep) (Amersham International, Amersham, Bucks). A working dilution of the HRP-conjugate solution is freshly prepared each day by adding $5\,\mu l$ to 4 ml of deionized water. The HRP-conjugate should have sufficient activity such that the addition of the Trolox standard solution gives a convenient t-trolox time of approximately 5 min (Figure 11.4A).

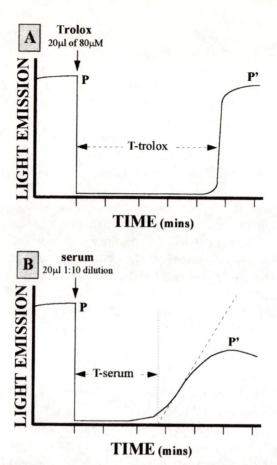

Figure 11.4   The kinetics of light emission from the ECL assay system. Light emission was measured in arbitrary units and time in minutes. A standard reaction was initiated by mixing of 900 μl deionized water, 100 μl Amerlite signal reagent and 20 μl HRP conjugate solution in a plastic cuvette, and gave an intense glowing light emission. (A) Following addition of 20 μl of 80 μM Trolox standard solution to the glowing reaction at peak intensity (P), light emission was immediately quenched. After a period of approximately 250 s, light emission resumed to reach a second peak (P'). The time taken to reach the intercept of the maximum slope of the recovery curve with the time axis was known as t-trolox. (B) Following addition of 20 μl of a 1:10 dilution of serum or plasma, light emission was immediately quenched. After a variable time period there was a gradual recovery of light emission. The period of complete light suppression was calculated as above and is known as t-serum. The total antioxidant activity of serum could be calculated from the two t-values using Equation 11.2. Serum from healthy volunteers produced typical t-serum values of between 100 and 150 s.

Calibration standard

Trolox (Sigma, Poole, Dorset). A standard antioxidant solution is prepared by dissolving 20 mg Trolox (mol. wt. 250.29) in PBS, diluting to 1 litre (80 μmol/l).

Apparatus

Light emission can be measured using a conventional lumin-ometer able to accept 1 ml cuvettes or tubes, such as a BioOrbit 1250 Luminometer (BioOrbit, Turku). Alternatively, it is possible to adapt other instruments with a photomultiplier tube such as a fluorimeter or spectrophotometer. Peak light emissions occur at 420 nm. Absolute calibration of the system is unnecessary and the intensity of light emission can be measured in arbitrary units since it is the emission *kinetics* rather than the *absolute* level that is important. The light emission kinetics are recorded on a chart recorder running at 0.5 mm/s.

Procedure

To initiate the ECL reaction, 900 μl deionized water, 100 μl of signal reagent and 20 μl of the diluted HRP-conjugate are added in that order to the reaction cuvette and carefully mixed. The cuvette is then placed in the luminometer and the reaction is allowed to 'run' until the light emission reading stabilizes at a peak value (P). At the same time as 20 μl of the test solution is added to the cuvette, the recording paper is marked and the cuvette, is then returned to the luminometer. The light output is completely suppressed if any chain-breaking antioxidants are present in the test solution (Figure 11.4B). Measurement is con-tinued until suppression of light emission ends and the output returns to a second peak (P'). The time of complete light suppres-sion (t) in seconds is represented by drawing a tangential line from the steepest point of the recovery curve to its intercept with the time axis. Plasma or serum samples should be diluted ten times for convenient measurement times (t-plasma). This time is compared with similar times measured after addition of various dilutions of the Trolox standard (Figure 11.4A).

Calculations

A linear standard curve can be created by plotting the t-trolox values for a variety of dilutions of the Trolox standard solution (80 μM) (Figure 11.5). The antioxidant activity (AOA) of a test plasma can then be determined by comparison of the period of light suppression it produces (t-plasma) with that produced by a standard solution (t-trolox) of concentration [trolox] by the equation:

$$AOA_{ECL} = \text{t-plasma} \times [\text{trolox}] \times 10/\text{t-trolox} \qquad (11.2)$$

The multiplication factor of 10 is used because plasma or serum samples have to be diluted 1:10 before a convenient comparison can be made with the standard solution. The results of this calculation are usually defined as μmol Trolox equivalents per litre (μM) rather than automatically assigning Trolox a stoichiometric value of 2.0.

Reference range

The normal reference range (mean $\pm$ SD) for AOA_{ECL} is 424 $\pm$ 86 μmol Trolox Eq./l. The within-batch and between-batch

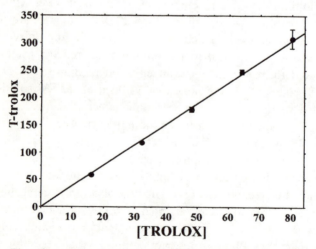

Figure 11.5 The relationship of t-trolox and Trolox concentration. In each case 20 μl of a Trolox solution of varying concentration (μM) was added to the ECL reaction and the delay to recovery of light emission (t-trolox) measured in seconds. Points are mean $\pm$ SD of 4 samples.

variability have been estimated to be 1.7% and 5.0% (CV), respectively. It has been calculated that the mean percentage contribution of the individual antioxidants to the total AOA_{ECL} is: urate 70%, ascorbate 10% and vitamin E 9%, but with little or no contribution from protein thiol groups. Heparinized plasma and serum give similar values but EDTA-plasma samples are 10% lower.

Interference

Any substances acting as a peroxidase will falsely accelerate the rate of the reaction, giving spuriously low t-values. Therefore, the measurement of haemolysed samples may be inaccurate. EDTA-plasma samples give values that are 10% less than corresponding heparinized plasma samples, which give similar results to serum. The light emission from turbid samples is attenuated but in all but extreme cases the kinetics of emission should remain visible.

Comments

The ECL assay offers the advantage of simplicity and precision. Its relatively high dependence on urate as an influence may be seen as a disadvantage. Its apparent lack of sensitivity to protein thiols may be seen as an advantage or disadvantage, depending on the circumstances. The need for a luminometer may pose problems for some laboratories, although alternative instrumentation can be adapted. As yet no adequate method of automation has been described. The ECL assay can also be used to measure antioxidant status in lipoprotein solutions where urate and other aqueous antioxidants have been removed by gel filtration (Maxwell *et al.*, 1994).

THE ABTS ASSAY

This assay is based on the observation that when 2,2′-azinobis (3-ethylbenzothiazoline 6-sulphonate) (ABTS) ($150\,\mu\text{M}$) is incubated with a peroxidase such as metmyoglobin ($2.5\,\mu\text{M}$) and H_2O_2 ($75\,\mu\text{M}$) the relatively long-lived

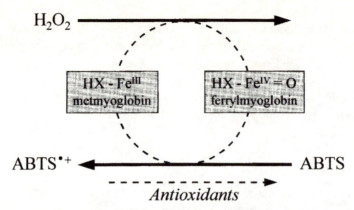

Figure 11.6   The formation of the ABTS$^{\bullet+}$ radical cation. The reaction of metmyo-globin with hydrogen peroxide forms the ferrylmyoglobin radical. This radical abstracts an electron from ABTS to form the blue-green ABTS$^{\bullet+}$ radical cation. The characteristic absorption of the radical cation can be inhibited by antioxidants and other reductants.

radical cation (ABTS$^{\bullet+}$) is formed (Figure 11.6). The ABTS$^{\bullet+}$ radical cation has a characteristic absorption spectrum with maxima at 650, 734 and 820 nm. In the presence of antioxidant reductants and hydrogen donors the absorbance of the ABTS$^{\bullet+}$ radical is quenched to an extent that is related to the antioxidant activity of the test solution (Miller *et al.*, 1993; Rice-Evans and Miller, 1994).

Reagents

1. Metmyoglobin (equine) (Sigma, St Louis, Missouri). Metmyoglobin should be purified prior to use on a 35 × 2.5 cm Sephadex G-15–120 column in PBS. The concentration of myoglobin in the eluate is calcul-ated from the extinction coefficients, and aliquots of metmyoglobin in PBS are stored frozen until use. A 400μM solution of metmyoglobin (0.0752 g in 10 ml of PBS) is prepared; 0.0244 g of potassium ferricyanide is dissolved in 100 ml PBS, and then 10 ml of the ferricyanide solution is added to 10 ml of metmyoglobin solution. The mixture is applied to the column and eluted with PBS. The first fraction is collected and its absorbance read at 490, 560, 580 and 700 nm. The absorbance reading at 700 nm is subtracted from the readings at 490, 560 and 580 nm to correct for background absorbance. Calculations of the relative propor-tions of the different forms of myoglobin can then be carried out using

deconvolution procedures or applying the Whitburn algorithms based on the extinction coefficients at 490, 550, and 580 nm:

$$[MetMb] = 146A_{490} - 108A_{560} + 2.1A_{580}$$
$$[FerrylMb] = -62A_{490} + 242A_{560} - 123A_{580}$$
$$[MbO_2] = 2.8A_{490} - 127A_{560} + 153A_{580}$$

where Mb is myoglobin. Myoglobin prepared in this way should only be used if it constitutes more than 94% of the total haem species present. The purified myoglobin is diluted to a concentration of 140 μM, divided into aliquots and frozen before use.

2. 2,2'-azinobis (3-ethylbenzothiazoline 6-sulphonic acid) diammonium salt (ABTS) (Aldrich, Milwaukee, Wisconsin). Prepare a 5 mM solution by dissolving 0.02743 g ABTS (mol. wt. 548.68) in 10.0 ml of PBS buffer. The extinction coefficient (ε_{mM}) of the resulting solution is 38.8 at 340 nm.
3. Hydrogen peroxide stock Aristar (BDH, Poole, Dorset). Dilute to a concentration of 500 mM in PBS by diluting 515 μl of Aristar H_2O_2 to 10 ml in PBS. This solution (45 μl) is further diluted to 50 ml with PBS to create a final concentration of 450 μM.
4. Phosphate-buffered saline (PBS), 5 mM, pH 7.4.

Standard solution

Trolox (Aldrich, Poole, Dorset). A stock 2.5 mM solution is prepared by dissolving 0.15641 g of Trolox (mol. wt. 250.29) in 250 ml PBS. Due to its poor solubility, Trolox may take time to dissolve and require mixing. Fresh working standards (0.5, 1.0, 1.5, 2.0 mM) are prepared daily.

Apparatus

A conventional spectrophotometer is required for the manual procedure and a Cobas Bio centrifugal analyser (Roche Diagnostic System Inc., Branchburg, New Jersey) is required for the automated assay.

Manual procedure

Mix 8.4 μl of sample, 489 μl of buffer, 36 μl of 70 μM metmyoglobin, and 300 μl of 5 mM ABTS to give a total volume of 1 ml, followed by vortex mixing.

The reaction is initiated by the addition of 167 μl of 450 μM H_2O_2, the clock started and the tube revortexed. This creates final reaction concentrations of 150 μM ABTS, 2.5 μM metmyoglobin, 75 μM H_2O_2 and 0.84% sample. The reaction mixture is transferred into a 1 cm cuvette in a spectrophotometer at 30°C. The absorbance of the samples and standards is read after incubation for 6 min at 30°C. This time is chosen because inhibition of the reaction by the top standard (2.5 mM Trolox) should have been overcome. There is a quantitative relationship between the absorbance at 734 nm at 6 min and the antioxidant activity of the sample or standard. Timing and temperature may be difficult to control within satifactory limits using the manual assay.

Automated procedure

The measurement protocol has been adapted for automated use on a Cobas Bio centrifugal analyser. ABTS/myoglobin reagent 300 μl is mixed with 30 μl of diluent and then 25 μl hydrogen peroxide is added as a starter to make a total incubation volume of 258 μl. The reaction is initiated by mixing of the reagents as a result of centrifugal forces. Final absorbance is read after 6 min of incubation at 30°C.

Calculations

The antioxidant activity of samples (AOA_{ABTS}) can be calculate by comparing its absorbance at 734 nm at 6 min with the calibration curve created by using the Trolox standards.

Reference range

The AOA_{ABTS} reference range (95th percentile) for adult human plasma has been reported as 1.32 to 1.60 mmol Trolox Eq./l (1.46 $\pm$ 0.14 mM). It has been calculated that the mean percentage contribution of the individual antioxidants to the total AOA_{ABTS} is: urate 33%, ascorbate 9%, vitamin E 3% and albumin 43%. The within-batch and between-batch variability have been estimated to be 1.6% and 6.1% (CV), respectively. Plasma samples can be collected using heparin, which does not interfere with the assay and gives similar values to serum. EDTA-plasma gives 10% lower values.

Interference

Substances acting as peroxidases, such as haem proteins in haemolysed samples, will promote the formation of the ABTS$^{\bullet+}$ radical cation. Any substances that display significant absorption at 734 nm might also cause positive interference.

Comments

The ABTS assay has the considerable advantage that it can be automated, enabling the measurement of large numbers of samples. It is partly limited by its relatively high dependence on protein concentration. This may explain the relatively small interindividual variability found amongst healthy adults (Table 11.3). The assay can also be used to measure antioxidant activity in lipoprotein solutions.

OTHER ANTIOXIDANT ASSAYS

The FRAP assay

This assay (Benzie and Strain, 1996) is based on the potential of antioxidants to act as electron donors and reduce ferric ions to ferrous ions in solution. This causes a colour change when the ferric ions are complexed with tripyr-idyltriazine at pH 3.6. The resulting ferrous complex has an intense blue colour, with absorption maximum at 593 nm. This process is convenient for a Cobas Fara centrifugal analyser.

The OPD assay

This assay (Nakamura *et al.*, 1987) employs *o*-phenylenediamine as a perox-idase substrate. When oxidized to its free radical form, OPD exhibits an increase in absorbance at 430 nm. In this assay, 500 μl of serum is mixed with OPD and H_2O_2 in citrate-phosphate buffer, pH 5.0. The change in absorb-ance is measured over a 2 h period.

The phycoerythrin assay

Peroxyl radicals are created as a result of the thermal decomposition of 2,2'-azobis(2-amidinopropane) (AAPH) at 37° (DeLange and Glazer, 1989). β-Phycoerythrin is normally highly fluorescent (excitation 540 nm; emission 565 nm) but loses this quality when damaged by peroxyl radicals. Therefore addition of samples containing antioxidants to the reaction mixture inhibits the loss of fluorescence intensity in proportion to the antioxidant activity in the sample. The original kinetic assay was subsequently modified to form the oxygen-radical absorbance capacity (ORAC) assay based on taking the reaction to completion (Cao *et al.*, 1993).

The *cis*-parinaric acid assay

cis-Parinaric acid is a lipophilic fluorophore (excitation 324 nm, emission 413 nm) that is destroyed on reaction with free radicals and this process can be inhibited by the presence of lipophilic antioxidants such as α-tocopherol (McKenna *et al.*, 1991). The oxidation is initiated by the addition of copper sulphate and cumene hydroperoxide and the loss of fluorescence is monitored over 30 min.

The plasma peroxidation potential assay

This assay (Arshad *et al.*, 1991) is based on the susceptibility of the lipoproteins in whole plasma to copper/H_2O_2-induced peroxidation. It necessitates the measurement of products of fatty acid and cholesterol oxidation by the thiobarbituric acid assay and gas–liquid chromatography, respectively.

REFERENCES

Arshad, M.A.Q., Bhadra, S., Cohen, R.M. and Subbiah, M.T.R. (1991) Plasma lipoprotein peroxidation potential: a test to evaluate individual susceptibility to oxidation. *Clin. Chem.*, **37**, 1756–1758.

Benzie, I.F.F. and Strain, J.J. (1996) The ferric reducing ability of plasma (FRAP) as a measure of antioxidant power: the FRAP assay. *Anal. Biochem.*, **239**, 70–76.

Cao, G., Alessio, H.M. and Cutler, R.G. (1993) Oxygen-radical absorbance capacity assay for antioxidants. *Free Radical Biol. Med.*, **14**, 303–311.

DeLange, R.J. and Glazer, A.N. (1989) Phycoerythrin fluorescence-based assay for peroxy radicals: a screen for biologically relevant protective agents. *Anal. Biochem.*, **177**, 300–306.

Goode, H.F., Richardson, N., Myers, D.S., Howdle, P.D., Walker, B.E. and Webster, N.R. (1995) The effect of anticoagulant choice on apparent total antioxidant capacity using three different methods. *Annals Clin. Biochem.*, **32**, 413–416.

Halliwell, B. (1990) How to characterize a biological antioxidant. *Free Radical Res. Commns.*, **9**, 1–32.

Halliwell, B. and Gutteridge, J.M.C. (1990) The antioxidants of human extracellular fluids. *Arch. Biochem. Biophys.*, **280**, 1–8.

Maxwell, S.R.J., Wiklund, O. and Bondjers, G. (1994) Measurement of antioxidant activity in lipoprotein fractions using enhanced chemiluminescence. *Atherosclerosis*, **111**, 79–89.

McKenna, R., Kezdy, F.J. and Epps, D.E. (1991) Kinetic-analysis of the free-radical-induced lipid-peroxidation in human membranes – erythrocyteuation of potential anvalidants using cis-parinaric acid to monitor perosidation. *Anal. Biochem.,* **196,** 443.

Metsä-Ketelä, T. (1991) Luminescent assay for total peroxyl radical-trapping capability of plasma. In: Stanley, P.E. and Kricka, L.J. (eds) Bioluminescence and chemiluminescence: current status, pp. 389–392. John Wiley & Sons, Chichester.

Metsä-Ketelä, T. and Kirkkola, A. L. (1992) *Free Radical Res. Commns.*, **16,** Suppl. 1: 215.

Miller, N.J., Rice-Evans, C., Davies, M.J., Gopinathan, V. and Milner, A. (1993) A novel method for measuring antioxidant capacity and its application to measuring the antioxidant status in premature neonates. *Clin. Sci.,* **84,** 407 412.

Nakamura, K., Endo, H. and Kashiwazaki, S. (1987) Serum oxidation activities and rheumatoid arthritis. *Int. J. Tissue Reaction,* **9,** 307–316.

Rice-Evans, C. and Miller, N.J. (1994) Total antioxidant status in plasma and body fluids. *Meth. Enzymol.,* **234,** 279–293.

Stocker, R. and Frei, B. (1991) Endogenous antioxidant defences in human blood plasma. In: Sies, H. (ed.) *Oxidative Stress: oxidants and antioxidants.* Academic Press, London, pp. 213–242.

Stocks, J., Gutteridge, J.M.C., Sharp, R.J. and Dormandy, T.L. (1974) Assay using brain homogenate for measuring antioxidant activity of biological fluids. *Clin. Sci. Mol. Med.,* **47,** 215–222.

Wayner, D.D.M., Burton, G.W., Ingold, K.U., Barclay, L.R.C. and Locke, S.J. (1987) The relative contributions of Vitamin E, urate, ascorbate and proteins to the total peroxyl radical-trapping antioxidant activity of human blood plasma. *Biochim. Biophys. Acta,* **924,** 408–419.

Whitehead, T.P., Thorpe, G.H.G. and Maxwell, S.R.J. (1992) An enhanced chemiluminescent assay for antioxidant capacity in biological fluids. *Anal. Chim. Acta,* **266,** 265–277.

Index